SCALES AND WEIGHTS

Yale Studies in the History of Science and Medicine, 1

SCALES AND WEIGHTS

A Historical Outline

by
BRUNO KISCH

New Haven and London, Yale University Press

Library of Congress catalog card number: 65–12545

*This book
is dedicated to the memory of
Professor John F. Fulton,
outstanding scientist and historian
and always kind, thoughtful, and
inspiring friend*

CONTENTS

Part II. Manufacture of Scales and Weights

LIST OF ILLUSTRATIONS

LIST OF CHARTS

LIST OF TABLES

PREFACE

There are two approaches to history. One is the collection, registration, and critical evaluation of ideas, the other a similar treatment of material objects from the past. The first results in publication of treatises and books, the second in collections and museums. Historical libraries represent both: collections of ideas committed to writing and collections of artifacts that have resulted from the ideas of their creators. These two approaches to history can never be entirely separated.

This monograph is devoted to the history of weighing; that is, to the idea of comparing as accurately as possible the mass of two objects. From ancient times reliable evaluation has been a prerequisite in the proper exchange of goods and therefore a basic necessity for commerce and for confidence and goodwill between men. Among the indispensable instruments serving this purpose are weights and balances, since all trade begins with counting, measuring, and weighing.

Weights and measures may also be studied for the ideas they represent. Masses of books have been written about weighing and measuring and the standards used at different times and in different places. The science of metrology deals with this problem. The work of Hultsch in the 1860s is still the classic source of information on the metrology of the ancient Greeks and Romans, although some of his conclusions have been superseded by the work of more recent scholars.

In the past, little attention was devoted to studying the development of weights and scales or to collecting them. Selected specimens, notable for appearance or age, have survived because of the interest of individuals, for the systematic collection of weights and measures was neglected by most museums until modern times. Even now such collecting is usually inspired by an interest in local history. Outstanding examples are the assemblage of old French weights in the Museum of Toulouse, with a masterly catalogue edited by Armand Machabey (1953), and the Greek weights in the Museum of Athens, still awaiting a thorough scientific study.

At the beginning of the last century, practical commercial considerations caused Lord Castlereagh, then Foreign Secretary, to collect the weights in current use in all countries with which England then (1818) had political affiliation. I was fortunate to find the greater part of this collection in perfect condition in the subterranean recesses of the Science Museum in London, and to be graciously permitted by Dr. C. St. C. Davison, the keeper, to examine it.

King Carlos IV of Spain had similar reasons in 1804 for bringing together a collection of weights from different countries. Although less extensive than that of Lord Castlereagh, it is still most valuable because, unlike Castlereagh's collection, it is composed of official standard weights; it is kept in the Museo Geographico in Madrid.

One of the most outstanding modern collections of weights (but mainly coin weights) is the one assembled by Gnecchi, in Milan. It includes 4,000 specimens and was acquired in 1926 by the Kunsthistorisches Museum in Vienna (Holzmair, 1932).

In the 1920s the late Dr. Edward Clark Streeter, with the intuition of the born collector, decided to preserve as many valuable metrological relics as he could find. The tangible result of his interest is a unique general accumulation of objects connected with weighing and measuring, together with examples of the pertinent literature, all of which he bequeathed to Yale University.

This volume is an outline of the history of weights and scales; a second volume, which is in preparation, is to be a catalogue of the weights and scales preserved in the Edward Clark Streeter Collection of Weights and Measures at Yale University, of which the author has the honor to be curator. The collection has been enlarged since Dr. Streeter's death by several important acquisitions, mainly of Greek, Roman, and medieval French weights, and a unique group of twenty-three large French commercial weights of the seventeenth, eighteenth, and nineteenth centuries.

It is a pleasure for the author to express his gratitude to all those who helped in this work. First, my late friend John F. Fulton, with constant interest and encouragement, urged me to continue and finally to finish this book. Professor Fulton's untimely death prevented my dedicating this volume to him personally, but it is dedicated to his memory in esteem and friendship.

My special gratitude goes also to the American Philosophical Society in Philadelphia. I was thrice privileged to receive grants from this organization, which enabled me to spend repeated summer vacations in Europe and the Near East, studying the treasures of the different museums and collecting data and material illustrative of the weights and scales of ancient times and distant places.

I am also deeply indebted to the directors of the museums and to the owners of private collections which I was permitted to visit, for their kindness, their helpful understanding, and their time. A list of the collections consulted for this book is included (see p. xix).

I wish to express my particular appreciation of the time and great effort of Madeline E. Stanton, librarian of the historical collections of the Yale Medical Library, in editing the English of this book and of Howard J. Reynolds, technical specialist in medical photography at the Yale School of Medicine, for making the excellent and often very difficult reproductions of the specimens from the Streeter Collection and my own collection.

B.K.

Brooklyn, New York
February 1964

COLLECTIONS AND ABBREVIATIONS

All collections listed here have been personally visited by the author except those preceded by *, of which only photographs were examined, or by ‡, of which only descriptions were available.

AO	Oudheidkundige Musea (Vleeshis) Antwerp
‡A.J	Collection, Prof. A. Jäger, Aachen
‡AaR	Collection, Richard von Rey, Aachen
AbR	Regional Museum of the Art Gallery of Aberdeen
BaW	Waagenmuseum, Balingen (Germany)
BeH	Heimatsmuseum, Bensberg (Germany)
BgG	Gruuthuse Museum, Bruges
*BKG	Formerly Staatliches Museum Berlin (Kunstgewerbemuseum)
BrH	Historisk Museum, Bergen
BsH	Historisches Museum, Basel
BsPH	Pharmaciehistorisches Museum, Basel
BtMu	Museum mĕsta Bratislavy (Czechoslovakia)
BxCMe	Collection de l'état Belge (Cabinet des Medailles), Brussels
BxRA	Musée royale d'Art et Histoire, Brussels
CoUl	Museum Unterlinden, Colmar (France)
CR	Rhätisches Museum, Chur (Switzerland)
CtYSt	Edward C. Streeter Collection, Yale University, New Haven
DHL	Hessisches Landesmuseum, Darmstadt
EMA	Museum of Antiquities, Edinburgh
FmK	Museum für Kunsthandwerk, Frankfurt am Main
FrW	Wetterauer Museum, Friedberg, Hesse
GAH	Musée d'Art et d'Histoire, Geneva
GöH	Göteborg Historiska Museet
*HaGL	Gustav-Lübcke-Museum, Hamm (Germany)
HdA	Apothekermuseum, Heidelberg
HdK	Kurpfälzisches Museum der Stadt Heidelberg
HeHM	Historisches Museum, Heilbronn
*HlMo	Staatliche Galerie, Moritzburg, Halle (Germany)
HoW	West-Friesch Museum, Hoorn (Netherlands)
JB	Bezallel Museum, Jerusalem
JHUA	Dept. of Archaeology, Hebrew University, Jerusalem
IT	Tiroler Volkskunstmuseum, Innsbruck

*KaB	Badisches Landesmuseum, Karlsruhe
KDN	Royal Collection of Coins and Medals, Danish National Museum, Copenhagen
KnF	Collection Frings, Cologne
KnFZ	Collection of the Forschungsinstitut für Zahnheilkunde, Cologne
KnGS	Geldgeschichtliche Sammlung der Kreissparkasse, Cologne
KnL	Collection of the Landeseichdirektion, Cologne
KnZ	Kolnisches Stadtmuseum im Zeughaus (formerly Rheinisches Museum), Cologne
‡KoMr	Mittelrhein-Museum der Stadt Koblenz (Germany)
‡KpK	Stadtisches Kramer-Museum, Kempen/Niederrhein (Germany)
KsK	Staatliche Kunstsammlung, Kassel
*LbH	Museen der Hansestadt, Lübeck
LeRN	Rijksmuseum voor de Geschiedenis der Natuurwetenschappen, Leiden
LSc	Science Museum, London
LScC	Collection of Lord Castlereagh in Science Museum, London
LuH	Luzerner Historisches Museum, Lucerne
LUP	Flinders Petrie Collection, University College, London
LWH	Wellcome Historical Medical Museum, London
MAN	Museo Arqueologico Nacional, Madrid
MG	Museo Geographico, Madrid
‡MGl	Städtisches Museum, Munich-Gladbach
*MsL	Landesmuseum für Kunst und Kulturgeschichte, Münster, Westphalia
MzA	Altertumsmuseum der Stadt Mainz (Germany)
*NeCS	Clemens-Sels-Museum, Neuss (Germany)
NGN	Germanisches Nationalmuseum, Nuremberg (Department of Scientific Instruments)
NH	Handelsmuseum, Nuremberg
‡NwK	Kreismuseum, Neuwied (Germany)
NYBK	Collection of Dr. and Mrs. Bruno Kisch, Brooklyn, New York
NYCD	Dale Collection, Columbia University, New York
NYNS	Museum of the American Numismatic Society, New York
PaH	Museum Hlavniho města Prahy (Czechoslovakia)
PBN	Bibliothèque Nationale (Coin and Medal Department), Paris
PCl	Musée Cluny, Paris
PCN	Conservatoire National des Arts et Métiers, Paris
PLa	Collection, M. François Lavague, Paris
PLu	Collection Lugan, Arles (May–July 1958, on exhibit in Musée Monetaire, Paris)

‡RbSt   Museum der Stadt Regensburg (Germany)
*SiS   Museum des Siegerlandes, Siegen (Germany)
SMk   Kgl. Mintkabinet, Stockholm
SN   Stockholm Nordiska Museet
‡SsK   Kulturhistorisches Museum, Stralsund (Germany)
StAD   Musée des Arts decoratives, Strasbourg (France)
StGH   Historisches Museum, St. Gallen (Switzerland)
*TrRh   Rheinisches Landesmuseum, Trier (Germany)
*UHe   Hellweg-Museum der Stadt Unna (Germany)
UUF   Uppsala Universitets Museum for Nordiska Fornsaker (Sweden)
VM   Museum im Vaduz (Liechtenstein)
WB   Bundessammlung von Medaillen, Münzen und Geldzeichen, Vienna
WiAu   Collection, Dr. Aussenbüttel, Witten-Annen, Westphalia (Germany)
*WSt   Museum der Stadt Worms (Germany)
‡WuSt   Städtisches Museum Wuppertal, Wuppertal-Elberfeld (Germany)
ZL   Schweizerisches Landesmuseum, Zürich
ZMI   Medicinhistorisches Institut, Zürich

PART I
Scales and Weights

1. INTRODUCTION

The peaceful transfer of goods from one owner to another requires the agreement of both concerning the value of the goods. Values as a rule are based on quality and desirability and are determined by the amount or the mass of the objects to be traded. Means for a reliable measure of these factors are the foundation of a just and ethical commerce.

Counting was undoubtedly the starting point of all commerce, but the ability to count apparently developed slowly among primitive men (as can still be demonstrated in some of the aboriginal cultures of Africa and Australia). Measuring was probably the next step in evaluating goods and experiences; here the process was to compare the size of parts of one's own body, like a foot or a cubit, with objects in the surrounding world. Aeons were to pass before man had progressed to the mystique of mathematics and the creation of precious time-measuring chronometers and complex measuring devices.

Like primitive counting and measuring, primitive weighing needed no special instruments. A rough judgment could be reached by taking an object in each hand to assess their relative weights. That this method was used up to the eighteenth century in certain branches of commerce is clear from Chambers' *Cyclopaedia* of 1728 (p. 28).

The use of measurements like a handful (*manipulus* or *fasciculus*), or a pinch (*pugillus*) or a spoonful, seemed satisfactory for early medical prescriptions or cookbooks for determining a certain amount of material which was neither particularly valuable nor potent (Rhodius, 1672; *Pharmacopoeia Matritensis*, 1762). The medical papyri of ancient Egypt give no proof of the use of weights in the pharmacies, even in the Eighteenth Dynasty, whereas gold, silver, copper, and lapis lazuli were weighed (Griffith, 1892, p. 436; Leake, 1952). Ridgeway (1892) tried to prove from the Iliad that the Greeks in Homer's time weighed gold (but not silver or base metal), and that they were already weighing wool. In the Bible the weighing of silver is repeatedly referred to, and the weight of the hair of King David's son Absalom is mentioned. A report of Hernan Cortes to Charles V mentions that the Mexicans in 1521 still sold everything by number and measure but never by weight (Guerra, 1960, p. 343).

Man's progress toward critical judgment led to the invention of weights, measures of capacity and length, and coined money. All these were important instruments of national and international commerce, and weights and

1

measures became important also for the natural scientists. One recalls the rapid development of chemistry, mainly through the work of Lavoisier in the eighteenth century, when scales and weights became an indispensable tool for every chemist. Unlike counting and measuring, however, weighing did not take even its most primitive standard units directly from the human body.

The history of weights and scales is dependent mainly on two sources: the actual surviving instruments and literary documents. Although of great importance in metrology, objects made according to a certain standard of weight (like coins) are of no help for our study, because even the best preserved of them will not indicate anything about the appearance of the weights and scales used for their adjustment. Not even the sensitivity of ancient scales can be judged from differences in the weight of well-preserved coins of the same issue, because many other factors may have contributed to the disparity.

The most reliable source of our knowledge about ancient weights and scales is the surviving objects themselves. Even fragments may bear important clues about the form, material, or mastermarks and hallmarks. The main drawback in collecting old weights and scales is the likelihood of error in regarding some smooth pebble or geometric object as a weight. Weigall (1908) has emphasized this for old Egyptian weights; it holds also for other groups. In museums I have seen smooth pebbles or hemispheres of glass with a fine patina exhibited as Etruscan weights, but among the fine collections of Etruscan art in the Museum of Orvieto I saw backgammon boards which had been excavated together with a supply of exactly the same kind of pebbles and hemispheres of glass.

An ancient weight cannot be accredited as such without a characterizing inscription or at least a typical form, like the Egyptian weight in the form of a cupcake, commonly used in the Late Kingdom. Of high importance, of course, is the weight of such an object if it is in perfect condition. The exact coincidence of its weight with a well-known standard weight of similar appearance will authenticate it. On the other hand a weight may be recognized as such by form and inscription even if it has not the correct weight, because wear and tear may have diminished it or oxidation (so common in lead weights) or incrustations may have increased it.

Literary documents, drawings, and sculptures are great sources of knowledge in this field. Most important, of course, are drawings and sculptures. One of the very old representations of a scale—a small balance which a man holds in his right hand—is contained in a relief from Hittite times (probably around 1000 B.C.) found in Karkemish and now at the Louvre (see pp. 27 ff.; Fig. 3).

Pictures reveal not only the use of scales and weights but also their form. For instance, the books of death found in sarcophagi in Egypt, and used since

2

the second millennium B.C., have very instructive pictures of scales and their use (Fig. 4). Greek scales and weighing scenes are found on Greek pottery (Fig. 6), and Ibel (1908) published a relief found in Pompeii depicting a butcher shop in which a scale of the steelyard type is hanging on the wall. A wall painting in Pompeii, in the Casa del Vetti, shows *amorini* working in the Roman mint. Here, as in Egyptian pictures, the scale with equal arms and two pans (the usual "balance") is used for testing the weight of the minted coins. This fresco is remarkable as an eyewitness report of the way the Roman mint was conducted (Sharp, 1909).

Since Egyptian and Roman times, scales and weights have been depicted in innumerable paintings in all countries, notably in pictures of St. Michael in the Christian era. As in the earlier Egyptian concept represented in the books of the dead, St. Michael was frequently shown weighing the soul of a deceased sinner. The Nuremberg relief of 1497, made by Adam Kraft (Fig. 15), is very impressive and instructive; it shows the use of a big scale and the form of the large commercial weights of the fifteenth century (Hartlaub, p. 157; Snyder, p. 13). Another extremely interesting scene is in a glass painting from the fifteenth century in a window of the cathedral of Tournay. Not only the weighing of a barrel on a big commercial scale can be seen but also the form and different hallmarks of the weights. (I owe the knowledge of this beautiful scene of medieval life to the kindness of H. H. Albrecht of Igls, Tirol.)

Strictly literary documents—for example, reports in old historiography (Herodotus and others) about tributes paid—are mainly of interest to metrologists, but occasionally they too give clues to the weights and scales as objects. When the Bible uses for weights the Hebrew expression *even* (stone) we learn (and the relics really prove it) that old Jewish weights, like those of Babylon, Egypt, and early Rome, were usually made from stone, not metal.

These documents, pictures, and objects provide information not only about the metrology of a people but also illuminate aspects of its cultural history and its economic and commercial status at a given time. Our concern in this monograph is limited, however, mainly to Central Europe, for otherwise it would have grown to unmanageable size.

Weighing is probably the youngest, least primitive of the three inventions of the human mind—counting, measuring, and weighing. It was also the last to be generally accepted and integrated, even in the field of natural science. A significant example of this is seen in the three-volume handbook of chemistry, written by the learned physician Johann Juncker; in the chapter devoted to the instruments of the chemist he does not mention weights and scales, and this was the middle of the eighteenth century.

Throughout Europe, governments early became aware of their duty to protect the citizens against being cheated in their everyday commerce, and

3

they turned their attention to supervising weights, scales, and money. It was probably the same all over the world wherever reasonable governments ruled. The coining of money soon became a government monopoly, and a similar monopoly of weights and measures may have been established (in the villages of France, for example) in the medieval period. At least the fact that the early French "monetiform" weights (Gaillardie, 1898) often bear the year of issue hints in this direction (see pp. 155 ff.).

Two problems were of constant concern to governments: to safeguard the proper manufacture and use of weights and measures, and to urge uniformity of standards, at least within their individual realms. They not only kept a watchful eye on the accuracy of weights and scales but they appointed experts as officers entrusted with the duty of checking the instruments. Among the oldest laws against the fraudulent use of weights and scales are the different Biblical commandments (Lev. 19: 35 ff.; Deut. 25: 14 ff.). Pretextatus, prefect of Rome, as early as A.D. 367 ordered uniformity of weights throughout the Empire (Beyerlinck, vol. 1, p. 88).

Charlemagne's "Admonitio generalis" of 789 (Küntzel, 1894) is well known and often quoted. He cautions all citizens "ut aequales mensuras et rectas et pondera justa et aequalia omnes habeant, sive in civitatibus sive in monasteriis" (all should have equal and correct measurements and just and equal weights in the cities and in the monasteries). A similar rule was established for the Bohemian kingdom by King Otokar in the thirteenth century (Beyerlinck, vol. 1, p. 88).

Along with the laws and ordinances issued by various rulers to maintain the honesty of commercial weights, scales, and measures came another important practice: coined money was guaranteed by the issuing government, and it was identified by its specific appearance. Forgery was severely punished, in early times as a rule by capital punishment. There is some reason to believe that local monopolies may have existed in relation to the monetiform weights in medieval France bearing the coat of arms of certain villages (Gaillardie) and perhaps also with the Greek weights which, like their coins, were marked with the symbols of towns or provinces. The market police then had only to check the weights being used, to prevent fraudulent manipulations such as reducing their weight by filing.

Wherever weight and scale making became a free enterprise, the instruments were checked from time to time by a government-appointed adjuster (*Aichmeister, ajusteur,* sealer) and were certified for use by a specific puncheon impressed on the object. Such a mark usually represented the coat of arms of a country or town, and its misuse brought heavy punishment. In the free city of Cologne, the adjuster's sign has always been the stem of Cologne (the three crowns), and no one except the sworn Aichmeister of the city was permitted to use the crowns for marking weights and scales. To prevent any mistakes, no weight maker in Cologne was permitted to choose a mastersign for himself in which a crown appeared (Kisch, 1960b). Official marking, well known from occasional Greek weights but subsequently generally neglected, was revived later in various European countries and the Near East (see p. 163). In the old Bohemian kingdom, for instance, King Otokar ruled in 1268 that all weights and measures should be replaced by instruments marked with his sign, a procedure unknown up to that time in his country (Sedláček, 1923, p. 161). Sedláček also reports (p. 154) that the generally accepted expression *mark* (in Czech, *marka*) for a half-pound weight stems from the practice of putting a certain mark on the weights as proof of their certification and gives the year 1042 as the first in which the use of this denomination can be proved. Adelung's dictionary of the German language (1788) gives a similar explanation for mark—as a coin or a weight unit that has been marked.

The history of governmental punching of weights would merit a separate study. That the Greeks did not invent this method of safeguarding the reliability of certain weights is proven by a Babylonian weight found in Syria, reported as No. 247 by Soloweitschik. It is inscribed: "Tested in the presence of the officers of the mint." But apparently these marks were not obligatory again for many centuries. We find them on some medieval Islamic bronze weights but often they are scarcely legible. The oldest such official mark on an Islamic bronze weight may be on one possessed by this writer. It bears the mark of a Fatimid Calif of Egypt who reigned from 1036 to 1094.* The weights in the form of little animals used in Scandinavia in the late Middle Ages (Fig. 81) also have certain marks that may be adjusters' signs (Kisch, 1959). However, among the many weights found in Viking graves, which I have had the opportunity to check, I found no mark that could reasonably be designated as a hallmark, governmental puncheon, or adjuster's sign. All this seems to prove that such marks became obligatory by governmental decrees only in the late medieval period in Europe.

The careful attention each government paid to the weights and scales used in commerce was not only for economic but also for religious reasons. According to the Bible it was, as already mentioned, a great sin to use wrong weights and measures. Quotations from the Bible on this subject are commonly found on the labels of Dutch gold-weighing scales in the seventeenth and eighteenth centuries; they are less common on German scales of that period.

Two fields of human activities closely connected with weighing and measuring attracted the special interest of governments: coinage and pharmaceutical dispensation.

---

* It was kindly deciphered for me by Prof. G. Miles of New York.

5

Coinage is always closely linked to weight. The coin (gold or silver) originally represented a governmental guarantee of the weight and value of that specific object. Later it was a guarantee of the government to exchange a certain object of metal or paper for the amount of gold or silver indicated on it. Early coins were of gold, electron, or silver in an amount rarely indicated by an inscription but designated by well-known laws and guaranteed by the governmental sign on obverse and reverse or on only one side, as in early Greek coinage. The relationship between coin and weight is still recalled by the names of certain coins which indicate their origin, like the "pound" in England (*pondus*, weight): the "mark" in Germany, as noted above; the "lira" (*libra*, pound) in Italy and modern Israel; the "shekel" in old Palestine and other countries of the Near East (*shokal*, to weigh). The names of the coins remain, but the coins have long since changed in standard and actual weight.

Coinage has always been checked at the mint, with very exact sets of weights, but the performance of this task was not handled too well in the old days and resulted (as all numismatists know) in a certain inequality in coins of the same type. Therefore in medieval Germany private persons were prohibited under severe penalty from possessing scales at all (Kisch, 1960b). The government thus tried to prevent its citizens from discovering and melting down overweight coins and reselling the metal to the mint with good profit.

The Roman *exagia* were officially checked coin weights, used in the mints in the classical period. Similar ones from France and Italy of the nineteenth century still exist. The correct weight of the silver or gold coin is given on them in grams up to the third decimal. In the nineteenth century these standards, like the old Roman exagia, were issued by the government to the mintmasters throughout the country.

In Greece and Rome (and probably also in ancient India) the original standard weights for comparison with the market weights were kept in the temples. The biblical expression "shekel of the sanctuary" may hint at this fact also. In the Roman Empire there were special buildings (*ponderaria*) in which the standard weights were kept. They were attended by a designated *ponderarius* (Pauly, vol. 22, p. 2425). In Cologne they were kept in the *Rentkammer*. Once a year the sworn Aichmeister had to compare with these *Muttergewichte* one of his two obligatory standard sets of weights; his second set had to be compared once a year with the adjusted first set. He used the second set throughout the year to check the weights brought to him by weight makers or other citizens for adjustment and to be marked with his official seal, bearing the three crowns, attesting their validity. In Paris the guild of pharmacists (*marchands, épiciers*) kept the standard weights. These

6

had to be compared every six years with the *métrices originales* kept in the royal mint (Felibien, 1725, vol. 2, p. 928).

Especially careful supervision of medical weights and scales was required. The physician originally was his own pharmacist, preparing and dispensing drugs to his patients. The pharmacist later took over this task under the supervision of the physician, and finally the doctor only made out the prescriptions to be filled by the pharmacist. Mistakes in the selection of weights and the use of badly adjusted weights or scales by physicians or pharmacists could endanger the life of a patient. The governments, therefore, paid particular attention to the instruments used in making up prescribed medicines.

2. EXACTNESS OF WEIGHING

Only during the last two centuries has weighing become scientifically exact. The discovery of the rare gases in the atmosphere (argon was the first) by Lord Rayleigh and Sir William Ramsay in 1896 began with their becoming aware in 1894 of discrepancies in the third decimal of Lord Rayleigh's findings and his theoretical expectations concerning the weight of "atmospheric nitrogen" (Rayleigh and Ramsay). Even in an ordinary chemical laboratory exact weighing is nowadays taken for granted in a manner never dreamed of in earlier days. To reach this goal, scales and weights had to be developed according to the highest scientific standards, and all students of chemistry and physics now must learn to use them properly.

Nevertheless, in some ancient countries scales and weights attained an astonishing degree of efficiency. Well-preserved stone weights of the same denomination from Egypt show a remarkable similarity despite almost certain deterioration in the millennia of their existence. In Roman times, as literary sources prove, there were weights of very small size like the siliqua (0.189 gm). Such a weight unit would have been impossible had there not been scales sensitive enough to indicate a value of less than 0.2 gram. In the Streeter Collection, sixteen one-kedet weights of diorite from the Late Kingdom of Egypt (ca. 1000 B.C.) have the following weights in grams: 8.55; 9.05; 9.05; 9.35; 9.25; 9.65; 9.55; 9.65; 9.95; 9.45; 9.55; 9.55; 9.65; 8.85; 8.95; 9.65. The arithmetical mean is therefore 9.356 grams. Since the age of these objects is about three thousand years, the general accord of their weight is astonishing indeed. In the Koran (21.47) Allah is quoted as saying that at the time of the resurrection he will use a scale so sensitive that it will indicate even the weight of a grain of mustard seed (Dietrich, 1954, p. 19).

Awareness of the importance of exact weighing apparently did not exist in Europe in most branches of commerce up to the end of the eighteenth century, and any earlier skill and interest had apparently been lost. In Arabia, however, the coinage of gold in this period was precise, as extant examples of ancient Islamic gold coins prove. The Arab people, well versed in astronomy, mathematics, and theoretical sciences, managed most problems involved in weighing with precision. Unbelievable as it sounds, Petrie (1918, p. 115) reports that Islamic coin weights of glass of the year A.D. 780 differ from each other not more than one third of a milligram (0.0003 gm). This proves the existence of extremely sensitive methods of weighing at an early date (Hinz, 1955, p. 1) and a sense of scientific responsibility among the mint-masters and balance makers, which medieval Europeans did not seem to have.

An instructive example of incompetence came to light in a report by a special committee to the Austrian Ministry of Commerce in 1870 (*Commissionsbericht*, pp. 102 ff.). The standard weight unit used for coinage in all parts of Germany and even in other countries since the middle ages has been the mark of Cologne (1 mark = ½ pound). The decree on coinage (*Münzordnung*) of Emperor Ferdinand I of August 1, 1560, stated that the Viennese mark, the weighing unit in Austrian countries, should have the ratio of five Viennese marks to six Cologne marks. This rule was of great importance, especially for gold coinage, and in the following centuries comparisons of these two standard mark weights were repeatedly made. Most important was a comparison made at the request of Emperor Leopold I, who ordered and received from Cologne in 1703 an exactly made standard weight of one mark, with which to check and justify the Viennese mark. Six "exact" copies of this Cologne mark were made in Vienna.

One hundred and seventy years later the Committee of 1870 (see *Commissionsbericht*) found these objects well preserved and most carefully stored. The mark weight of Cologne was 243 milligrams heavier than it should have been. Three of the original standard weights of Cologne, each supposed to weigh one pound, were also extant in 1870 and, according to the report of 1703 given to Emperor Leopold I, were "exactly even in weight." They were checked by the Committee of 1870 and found to weigh 467.548, 467.737, and 468.125 grams. The difference between the heaviest and the lightest of these three *exactly equal* standard weights of one pound was more than half a gram (577 mg).

The five members of the Committee of 1870 reached the conclusion that in the sixteenth century a one-pound object could not be weighed with less than an error of 0.5 gram. The sensitivity of balances was not to be blamed for this inexact "exactness" because there were weights of much smaller denomination, which proved that the small scales used by money changers and pharmacists had a much higher sensitivity than half a gram.

Despite this lack of exactitude, it must be borne in mind that governments had always tried to maintain reliable weights chiefly because of the coinage. At a time when gold and silver were scarce, exact weighing was highly important. In France an edict of 1540 by Francis I stated the permissible margin of error for a weight of 25 marks (12.5 pounds); it was set at 1.5 esterlin (1 esterlin = 1/160 of a mark). The margin of error for a nested weight of 8 marks was 3 felins, for 1 mark 0.5 felin (1 felin = 1/640 of a mark) (Eisenschmid, 1737, p. 2). In Cologne, since medieval times there have been laws

(Kisch, 1960b) to regulate the making of exact weights and measures. They are summarized in 1553 in the constitution of the guild of weight and balance makers.

STANDARDS FOR WEIGHING

Like every other sophisticated activity of mankind, weighing and measuring have their own basic pattern. The pattern has changed somewhat during historical times, but the attitude that weighing should lead to a reliable statement remains. Weights and scales have therefore had to be progressively improved with the increasing demand, mainly from scientists, for exactness in weighing data.

At all times, emphasis has been placed mainly on weighing valuable things or, in the medical realm, the potent drugs. Glanville's research (1935) has indicated that in Egypt at the time of the Old and Middle Kingdoms (ca. 3000 to 1800 B.C.) weights and scales were used only by high officials of the king to weigh gold, silver, copper, and other precious material. In the marketplace of that time scales were not needed because goods were exchanged by barter. Glanville refers to the few pictures of market scenes found in Egyptian tombs of this period, where an exchange of goods can be seen but there is no evidence of scales. Not until the Twentieth and Twenty-first Dynasties (1200 to 950 B.C.) do pictures show the use of hand scales by private persons in the market. Weights were not used in ancient Egypt even for pharmaceutical preparations (Griffith, 1892, p. 436).

Olaus Magnus, Archbishop of Uppsala, reports in his remarkable book on the history of the Scandinavian people (1555, p. 468) that some of them ("Bothnienses sive gens Lapporum") even in his time (the sixteenth century), are "ignorant of figures, weights, and measures and live by exchange of goods." Emphasizing the predominance of the barter system he also stated (p. 444) that in the nordic countries weights were nonexistent or rare. The same was said of the Mexicans in 1512 by Hernan Cortes (Guerra, 1960, p. 343).

According to Étienne Boileau, who in 1258 was provost of Paris, a distinction was made there between merchants who did not sell by weight and those who did. The latter were subject to inspection (Depping, 1837, p. 440). In France the butchers offered the king in 1368 a certain amount of silver and an annual revenue if he would abrogate the custom of selling meat by weight (Clémenceau, 1909, p. 51).

Even today, not all kinds of goods are weighed in the markets. Nowhere are eggs sold by weight in spite of the fact that such a procedure would be reasonable. Fruit is often sold by the bushel, and no one is too careful to see

10

whether a bushel contains an apple more or less, and carrots and broccoli are sold by the bunch. But gold and silver have always been traded by weight, and jewels are carefully weighed. Herbs and teas to make medicinal infusions were measured by a "handful" or a "pinch" and are still used in rather inexact amounts, but potent drugs like atropine or adrenaline are today prescribed and dispensed according to their weight, and this weight is determined as exactly as possible.

The modern approach to measurements is very different from that of olden times. Today the standard for every weight and measure is a unit arbitrarily set by the government, be it gram or ounce, or meter or yard. The value of goods bought or sold is determined by this unit. One kilogram of whatever we are buying, whether it is salt or gold or fruit, is always exactly 1,000 grams of the accepted unit of the metric-decimal system. The weights used to determine this amount are standardized and marked and must be used by the merchant for every kind of goods. This philosophy is not very old. The concept of ethical weighing formerly involved different standard weights for different kinds of goods. We know that different standards of weights existed in Babylonia; at least there was a royal one and a common one (see Viedebantt, 1917). The royal standard was of course heavier, probably to prevent the king's treasury from suffering any loss from inexact weighing in the payment of tributes or duties. Even in the Bible the shekel of the sanctuary was apparently heavier than the common shekel, 3,000 shekels being one talent (*kikar*) (Exod. 38). But this expression could also mean that a standard shekel was kept in the sanctuary. This is mentioned here only to suggest some of the problems that arise when surviving weights are compared with standards known from ancient literature.

The Romans were probably the first to abandon the custom of having different weights for different goods, which had apparently been a popular practice also in classical Greece. Paetus writes in 1573 (p. 43): "Cum autem diversis gentibus · et presertim graecis diversarum rerum diversa pondera fuerint, Romanis tamen rerum omnium unum pondus existit nempe libra" (Whereas different people, especially the Greeks, had different weights for different goods, the Romans had for all things only one weight unit: the libra). But in medieval times this logical Roman attitude had been abandoned, not only in Europe but also in the Near East. Levy (1938, p. 26) quotes an Arab author of the fourteenth century who emphasized that each district not only had a different ratl (Arab weight; see p. 222) but that the *muhtasib* (the adjuster) had to know the different ratls that were used for different commodities.

The centner or quintal (the old Roman *centipondium* or *centenarius*) was invented, according to St. Epiphanias, by the Romans (Dean, 1935). It remained a much-used weighing unit up to modern times. The name means, literally, one hundred pounds, but in Europe from the Middle Ages to the nineteenth century a hundredweight for iron ore, for example, had a weight very different from that used to weigh butter or silk, even in the same country.

The old government regulations of Cologne give considerable information about conditions there. We know that in the fourteenth century the weight of a centner in the *Garnhaus* (the marketplace for yarn) was 106 pounds, but a centner for the iron scale weighed 120 pounds, that for the silk scale was 100 pounds, and the centner for general use weighed 104 pounds (Kisch, 1960b). A city scale was set up in Frankfurt am Main in 1558. The centner for spices was $109\frac{5}{16}$ pounds, and the centner for lard $117\frac{9}{16}$ pounds silver weight, but the merchant in the market used a centner of 108 pounds (Chelius, 1805, p. 77). All goods weighing over one fourth of a centner had to be weighed by a government officer on the public scales—that is, no merchant was permitted to weigh on his own scales anything heavier than 25 pounds (Kisch, 1960b). This law safeguarded an important income for the government. A similar law obtained in France (Paris and environs, for instance) from 1322 to the French Revolution (law of March 15–28, 1790). The centenarii (hundred-pounders) were made exclusively for use in governmental weighing places and were named according to the ruler: *poids-le-roi, poids seigneuriaux, poids-de-ville* (Testut, 1946, p. 45). In Germany the public scales were called *Stadtwaage, Güterwaag*, or according to the place where they stood: in Cologne, *Altermarkt, Kaufhaus Gürzenich*; or according to their use: *Kirschenwaage, Seidenwaage*, or in Mainz, *Goldwaage*.

In Kutna Hora, Czechoslovakia, there is an old metal centenarius, with a Czech inscription indicating that it is a centner, which weighs 120 pounds. There was another in the city of Německy Brod (in Bohemia), also of metal, which weighed 61 kilograms and 650 grams in 1917, i.e. more than 123 (metric) pounds (Sedláček).

Georg Agricola, the great scientist and physician of the sixteenth century, explained why the hundred-pound weight for iron ore weighed 120 pounds: some of the valuable metal remained in the slag in the melting process; it was therefore just to give the buyer an additional 20 per cent.

Another example of this philosophy of weighing comes from Aleppo, once a famous commercial center in Syria, where the unit of weight was the *rotto* (*rottolo*). Three rottolos of different weight were still in use there in the nineteenth century (Gregory, 1822). To weigh cotton, galls, and large commodities a rottolo of 720 drams was used; to weigh various silks, a rottolo of 624 drams was used; but the rottolo to weigh white silk was one of 700 drams.

Adjusting the weight units used for different kinds of goods used to be considered as reasonable and ethical as adjusting prices. This explains why

12

cities had different scales for different merchandise, managed by the government and officially denominated according to their purpose. In Cologne the government maintained different public scales (each supervised by special officers) for fat, silk, iron, herbs (for pharmacies, the so-called *Krautwaage*), and even a special cherry scale to weigh baskets of imported cherries. In Mainz there was a governmental gold scale in medieval times, and in Paris in the eighteenth century there was a scale for human beings who wanted to keep track of their personal weight and its changes for medical reasons; this scale was operated by a royal weighing official (Kisch, 1960a).

In France up to the time of Louis VII (1126?–80) the king was the owner of two public scales in Paris. The one for general uses was called the king's scale (*poids-le-roy*); the other was used only for weighing wax (*poids de la cire*) and was kept in a house called "le poids de la chandellerie" (Felibien, 1725, vol. 1, p. 198). Candles and candlemaking were most important in those days, and special scales were used for wax and for candles in Cologne and Paris.

In Cologne, by order of the government, a certain set of weights was kept with each public scale. The number of weights and the material from which they were made (lead or copper) were exactly prescribed. For each balance the weight of the pertinent centner was regulated, but none of the centipondia of Cologne except the one used at the silk scale had a standard of 100 pounds.

A current relic of the time-honored system of weighing different goods with different weights and different metrological standards in the same country is the simultaneous use of troy weight and avoirdupois weight in the modern Anglo-Saxon world (175 pounds troy = 144 pounds avoirdupois). Harris says of this practice in his *Lexicon Technicum* of 1723 (vol. 2, no pagination):

> Weights in use in England: One called Troy-Weight having 12 ounces in the pound, and by this jewels, silver, Gold, Corn, Bread and all Liquors are usually weighed and the other is called Averdupois, containing 16 ounces in the Pound: by this all coarse, drossy wastable wares such as Grocery, Pitch, Tar, Rosin, Wax, Tallow, Copper, Tin, Lead, Iron etc. are weighed.

All pharmaceutical operations in the Anglo-Saxon countries use the troy weight. Gregory reported (1822) that with the avoirdupois pound (453.25 gm) are weighed: mercury, groceries, base metals, wool, tallow, hemp, drugs, bread, etc.; and with the troy pound (372.96 gm) gold, silver, jewels, liquors, and pharmaceuticals.

Still more detrimental for the ethical exchange of goods in daily commerce than the coexistence of troy and avoirdupois weights was the fact that

13

in pre-metric times in most countries every village and marketplace had its own standard of weights. Innumerable rules tried to stop this confusion, with little avail. The decree of Charlemagne in A.D. 789 and the laws of King Otokar of Bohemia in the thirteenth century have already been mentioned (p. 4).

The Magna Charta (1215) stressed the point that uniformity of weights and measures should exist in England. However, even in 1794 Martin was deploring the "difficulties arising" to men of trade and commerce from the variety of "Weights and Measures used in different parts of England."

If conditions were in a desperate state in England, they were still worse in Germany with all its small countries and independent cities and in France where, as we know from an edict of King Henry II (1557), many different rulers enjoyed the right to determine the standard of weights and measures in their realms. The edict mentions (Garrault, 1585, p. 6): "des Princes, Prelates, Ducs, Marquis, Contes, Vicontes, Barrons, Chastellains et autres ayas droict de poids et mesures." This metrological chaos was also reflected in international commerce. Diderot's *Encyclopédie* summarizes the status at the end of the eighteenth century (1765 ed., vol. 12, p. 855) with these words: "Chaque pays a ses poids different non-seulement en Europe, mais dan les échelles du levant, en Asie, en Afrique etc." Chambers in his *Cyclopaedia* says in despair (vol. 2, p. 36): "This diversity of weights makes one of the most perplexing articles in commerce; but it is irremediable."

It can be mentioned in passing that using different weight units for weighing different materials, now entirely abandoned except in the Anglo-Saxon countries, creates great difficulty for the modern metrologist if he tries to set up the metrologic system of a bygone era according to the preserved relics (see the catalogue of weights in Appendix 3). This fact accounts, at least in part, for many controversies among outstanding authorities on the standards of weights of ancient peoples like the Babylonians and Assyrians.

The miraculous event at the end of the eighteenth century that brought order to the chaos of weights and measures was the development of the metric-decimal system. This inaugurated an entirely new era in weights and scales.

3. THE METRIC-DECIMAL SYSTEM

Aurum probatur igni ingenium vero mathematicis—Rivius, 1547

John Quincy Adams, later the sixth President of the United States, reminds the reader of his book on the metric system that the establishment of the French Republic had taken place on September 22, 1792, the day of the autumnal equinox, when the sun entered the sign of the balance, the symbol of equity. It was significant of the spirit of the French Revolution that in spite of the tumult, and in the midst of the excitement, cruelty, and momentous events, there was still the inclination to approach scientific and cultural problems from an entirely new viewpoint.

The introduction of the metric-decimal system was unquestionably among the greatest cultural contributions of the French Revolution. After a transition period of a few decades of vacillation between the popular, familiar, and impractical local systems of weighing and measuring and the strange new concept, the latter was made compulsory in France on January 1, 1840. It has subsequently been accepted by the greater part of the civilized world. This unique system, which we owe to French ingenuity and to the energy of its revolutionary forces, brought to an end the uncertainty caused by metrological disparities.

The new metric-decimal concept was a child of the rationalism of the times, and it involved the solution of two problems, equally important. Weights and measures and their deplorable lack of standardization were among the first problems to be tackled by the Assemblée Nationale in 1790. A new natural science was to be created in the midst of the political upheaval, but a standard for weights and measures, which was immutable, had first to be found. It was to be chosen from an entity in nature independent of tradition or the arbitrary decision of any king or ruler, and such that no nation would reject it out of pride or envy or for some similar subjective reason. The decision finally reached by French scientists and endorsed by the French government was the length of the meridian from Dunkirk to Barcelona. This would be newly measured and provide a standard unit—the meter—one forty-millionth part of a meridian of the earth.

The second problem (and most important for science and daily life) was to obtain general acceptance of the system of decadic division for all metrology.

15

The new meter seemed an outstandingly acceptable unit, but basically it made no great difference whether the "mètre," or the "mètre des Archives," or the "kilogramme des Archives" was to become the standard, or whether a yard or foot or mark was accepted as such, if only every country adopted it.

Weight reforms had earlier claimed the attention of many people. Diderot's *Encyclopédie*, in the 1765 edition (vol. 12, p. 855), repeated the complaint made in Chambers' *Cyclopaedia* of 1728 about the impossibility of finding a remedy for the desolate diversity of weights: "La diversité des poids fait un des articles des plus embarrassans dans le commerce, mais c'est un inconvénient irrémédiable." This paragraph is a literal translation of the statement of Chambers, whose *Cyclopaedia* was a most important incentive for Diderot and the encyclopedists. "Non-seulement la réduction des poids de toutes les nations à un seul est une chose impossible," the *Encyclopédie* continues, "mais la réduction même des différents poids établis dans une seule nation n'est pas practicable, témoin les efforts inutiles qu'on a faits en France pour réduire les poids sous Charlemagne, Philippe-le-Long, Louis XI, François I, Henri II, Charles IX, Henri III, Louis XIV."

In the beginning of the eighteenth century, in England as well as in France, the need to reform the metrological differences between these two countries was recognized. The discussions carried on in the *Philosophical Transactions* of the Royal Society of London (1742, *42*, 185; 1743, *43*, 541) and in the *Comptes rendus* of the Academy of Sciences in Paris give proof of the endeavors to reconcile the differences in their standards (Desaguliers, 1720). They also contain valuable information concerning the whereabouts and the kinds of standard weights and balances in these countries (Harkness, 1887). Even earlier (1670) Gabriel Monton, vicar of St. Paul's Church in Lyons, had proposed a decimal system based on the length of the arc subtended by an angle of one minute of a great circle of the earth (Research Report, 42, p. 9).

Picard in 1671 and Huygens in 1673 had suggested the introduction of a unit of measurement that was based on the length of a one-second pendulum. This idea was again stressed by La Condamine in Paris, together with the suggestion of a decimal system (p. 513), and later (1790) by the statesman Talleyrand, who suggested as a unit of measurement a standard based on the length of a pendulum beating seconds on latitude 45° (Kunz, 1913).

THE STANDARD FOR METRIC-DECIMAL WEIGHTS

There is good reason to suppose (Berriman, 1953, p. 6 ff.) that the standard of the Babylonian weight unit was based upon the weight of a cubic inch of gold. Berriman has tried to prove this theory by means of old, well-preserved weights from Babylon and by considering a collection of copper ingots in the Museum of Athens, seventeen of which were found in the sea off the east coast of Euboea. The weight unit seems in this case to be again a cubic inch of gold weighing 315 grams. Berriman also supposed that the Babylonian talent of 60 livres would be equal in mass to a Greek cubic foot of water at maximum density. This weight standard, with slight variations, turned up again in the metrology of Egypt, Greece, and Rome and, after the downfall of the Roman Empire, not only in Byzantium but all over Europe in the units of the pound with its different denominations (*libra, litra*; in Islamic countries, *ratl, rottolo*, and so on). In spite of the differences in weight units used in different countries—and even in the same country when, for instance, the French troy pound was used simultaneously with the avoirdupois in Great Britain—the variations in weight were not great enough to let us forget or overlook their common origin.

When France decided to adopt a metric-decimal metrology, a standard unit of weight had to be found for this system. At the beginning of this development (1788) Lavoisier and Haüy were ordered to determine the exact weight of 1 cubic decimeter of distilled water at freezing point weighed in a vacuum. This unit was then to be known as a *kilogramme provisoire* (Morin, 1873, p. 18). Lavoisier created the instruments of the required great exactitude to make this determination, and a committee consisting of Borda, Lagrange, Laplace, Monge, and Condorcet reported on the experiments to the Academy of Paris on March 19, 1791. The first kilogram weight was made of pewter and weighed 18,841 grains in air. Similar determinations of the weight of a certain volume of water were later made by Lefèvre-Guineau and by Trallés in 1796 (Morin, p. 19), who compared the weight at different temperatures (and accordingly at different densities) and also compared the weight in air with that in a vacuum. In 1795, a specialist in metallurgic problems, Jeannetti in Marseilles, was commissioned to make a meter and a kilogram of platinum. Jeannetti delivered four kilograms, three of them in 1801 (Wolf, 1882, p. 67).

Wolf's report on the standard of weights also gave an account of the further developments: the *kilogramme définitive* was made under the supervision of Lefèvre-Guineau in the workshop of Fortin in Paris, and Wolf gives exact measurements of the weight. The shop of this master was first in the rue St. Honoré, then on the Place de la Sorbonne, then on the rue de la Montaigne St. Génévière, and finally at the École Centrale du Panthéon. Fortin was also the master who checked the weight of the four kilograms of platinum delivered by Jeannetti; he found one of them not satisfactory and returned it to the maker. There is a further discussion of the material of the standard kilogram of France and other countries in Chapter 5 on page 79 ff.

Today's international definition of the kilogram no longer refers to a certain volume of water (1 cubic decimeter) but describes the kilogram as a standard unit of mass that is equal to the international prototype, the *kilogramme des Archives*, which is kept in France.

THE HISTORY OF THE METRIC SYSTEM

A fine historical outline of the entire development of the "systeme metrique" was presented by John Quincy Adams, Secretary of State of the United States, in a report prepared in compliance with a resolution of the Senate of March 3, 1817, and published in 1820. The extensive report of Morin,* published anonymously in 1873, is another excellent résumé. It depicts impressively the different phases in the introduction of the metric-decimal system. Instructive historical surveys of the metric system had previously been given (1869) by Dumas and by Chevreul at meetings of the Paris Academy of Science.

Later publications have added no significant data, but the more recent historical outline of Block (1928) is worth mentioning. Highlights in the development of the metric-decimal system are:

1790 M. de Talleyrand suggested to the Assemblée Nationale the creation of a "système uniforme" of weights and measures which would be acceptable to enlightened nations of the world. The suggestion was accepted by the Assemblée in an order of May 8, 1790, and confirmed by King Louis XVI on August 22, 1790. The decree provided for collaboration of the scientific academies of Paris and London in this matter. Still considered as a basis for the new standard was the length of a one-second pendulum determined at a latitude of 45 degrees, or some other basis still to be determined by the learned academies.

1792 In agreement with a committee report of March 19, 1791 (members: Borda, Condorcet, Lagrange, Laplace, Monge), the Assemblée Nationale ordered on March 26, 1792, that the measurement of the meridian from Dunkirk to Barcelona be started. Two scientists, Méchain and Delambre, were entrusted with this job. Their romantic and dangerous adventures during their travels among superstitious and illiterate populations in Jacobin-infested countries would make an exciting film.

1793 On August 1, 1793, the standard of a "mètre provisoire" was created by law to be valid to 1799. The decimal system was accepted as well.

1795 The National Convention decided on the definite nomenclature of the new weights and measures (April 7).

1799 A law of December, 1799, stated the standard values of weights and measures of France. The mètre provisoire and the kilogramme provisoire of 1793 were abolished. A specimen of the definitive meter and of the kilogram (made of platinum) had already been deposited on June 22, 1799, with the "Corps legislatif" at the Institut National des Sciences et des Arts. Exact copies of both were to be made and used as standards for the new weights and measures throughout the country. By the same law a medal was

---

* General A. Morin, member of the Institut National, was "Directeur du Conservatoir Imperial des Arts et des Métiers."

ordered to be struck to commemorate this event. This decree was signed by the three consuls: Siéyès, Bonaparte, and Ducos.

The platinum standard specimens were finally deposited in the Archives.

These decreed weights and measures could not suddenly replace the familiar old ones; wholehearted support of the new system was not given, nor was enforcement possible. Even Napoleon seems not to have grasped the overwhelming importance of this innovation.

1837 On July 4 a final decision was made by King Louis-Philippe: the metric-decimal system should become the only legally permissible standard system of weights and measures for all of France.

1840 January 1 was set for inauguration of the above decree.

1869 Shortly before the outbreak of the Franco-Prussian War, by decree of Napoleon III (September 1), "La Commission Internationale du Mètre" was founded. The French Government invited all foreign countries to send delegates to this convention.

1870 The first meeting of the Commission was scheduled for August 1, but because of war had to be postponed.

1872 Not until two years later, on September 24, did the meeting actually take place in Paris.

1927 France defines the meter as 1,553,164.13 wavelengths of cadmium-red light.

Resistance against the new system was great everywhere, chiefly owing to mental inertia, but partly because of psychological difficulties and the expense of changing all weights and measures on a national scale. Political considerations and prejudice were also involved; there were strong feelings against the revolutionaries, as antiroyalists and belligerent enemies. These sentiments were especially strong among the conservative populations of England and the rest of the Anglo-Saxon world, as was made clear by Sir George Shuckburgh Evelyn in the *Philosophical Transactions* of the Royal Society in London (1798, p. 165), when he sarcastically discussed the French proposals:

some more magnificent integer than the English pound or fathom: such as the diameter or circumference of the world, etc. etc. and without much skill in the learned languages and with little difficulty I might ape the barbarisms of the present day. But in truth, with much inconvenience, I see no possible good in changing the quantities, the divisions or the names of things of such constant recurrence in common life. I should therefore humbly submit it to the good sense of the people of these kingdoms at least, to preserve, with the measures, the language of their forefathers. I would call a yard a yard and a pound a pound, without any other alteration than what the precision of our own artists may obtain for us, or what the lapse of ages, or the teeth of time, may have required.

England and the United States were reluctant to compel acceptance of the metric-decimal system. An enormous literature has since been published in

19

these countries defending this position, some of it even reflecting the attitude of Evelyn.

Despite the fact that Jefferson had suggested in 1799 the acceptance of a decimal system (Taylor, 1887), John Quincy Adams came to the unfortunate conclusion that a compulsory acceptance of it by the United States would not be advisable. He arrived at this decision mainly out of practical considerations, but he also wrote:

> This system approaches to the ideal perfection of uniformity applied to weights and measures, and whether destined to succeed, or doomed to fail, will shed unfading glory upon the age in which it was conceived and upon the nation by which its execution was attempted and has been in part achieved.

Among the many publications in the United States seeking to justify the rejection of the metric-decimal system in this country, only a few can be mentioned. They defend the old system despite the fact that the Congress on July 28, 1866, had already made the use of the metric system permissive. Some of the arguments may be found in Halsey's book, *The Metric Fallacy* (1920), and in the National Industrial Conference's Research Report No. 42 (1921) under the title *The Metric Versus the English System of Weights and Measures*, and in the book of W. R. Ingalls, President of the American Institute of Weights and Measures, entitled *Modern Weights and Measures*. A bimonthly, *The International Standard*, may be mentioned as a curiosity. This journal was published by an "International Institute for Preserving and Perfecting Weights and Measures" in Boston and Cleveland in 1883. In a kind of editorial by Charles Latimer it was stated that this society had been founded at "noon of the eighth day of November 1879, when at the Old South Church, Boston—that spot of wonderful memories and more wonderful prediction—we asked the blessing of God upon our undertaking and the guidance of His Holy Spirit through the merits of His Son." On page 2 Latimer continues: "The issue forced upon us today is either the adoption of the French unit of measure, born of infidelity and atheism in 1792 . . ."

Exceptions to the defenders of the system of troy and avoirdupois were J. Pickering Putnam (1877) and more particularly the president of Columbia College, Frederick A. P. Barnard (1872). With admirable eloquence Barnard strongly recommended the adoption of the metric-decimal system in the United States. His brilliant endeavor was not a practical success, but the metric system has come to be more and more accepted in the Anglo-Saxon world in the field of the sciences, where it is a prerequisite for international exchange of information.

All but the Anglo-Saxon world soon turned to the new standards. The sequence, according to Research Report No. 42 was as follows:

Argentina	1863	Japan	1891
Austria	1873	Latvia	1919
Belgium	1816	Luxembourg	1816
Bolivia	1870	Mexico	1862
Bosnia	1878	Monaco	1876
Brazil	1862	Montenegro	1888
Bulgaria	1888	Netherlands	1816
Canada (permissible)	1871	Norway	1879
Chile	1848	Panama	1853
China	1914	Peru	1863
Colombia	1853	Poland	1919
Congo	1910	Portugal	1852
Cuba	1849	Puerto Rico	1849
Denmark	1910	Romania	1884
Ecuador	1856	Russia (permissible)	1900
Egypt (permissible)	1873	(compulsory)	1919
Finland	1886	San Domingo	1867
France	1840	Serbia	1873
Germany (permissible)	1870	Siam	1889
(compulsory)	1872	Spain	1849
Prussia (compulsory for medicinal		Sweden	1879
purposes)	1867	Switzerland	1875
Great Britain (permissible)	1864	Tripoli	1913
Greece	1849	Tunis	1895
Herzegovina	1878	Turkey	1886
Hungary	1876	United States (permissible)	1866
India (permissible)	1871	Uraguay	1862
(compulsory)	1960	Venezuela	1857
Italy	1845		

The triumph of the metric-decimal system in commerce, science, and medicine has justified the enthusiasm with which the three consuls of France had ordered the inscription for the medal commemorating the final determination of the standard for weights and measures in France by the law of the 19th of Frimaire, the year VIII (December 10, 1799). The inscription reads: "À tous les temps, à tous les peuples." This bronze medal (diameter: 70 mm), the picture of which (Fig. 1) I owe to the great kindness of Dr. A. Machabey, Chef du Service de Documentation et d'Etudes in Paris, was not actually struck until 1840, the year when the metric system became compulsory in France. According to Dr. Machabey it was "un amateur lyonais Pierre-Marie Gonon qui fit graver en 1840 par son compatriote Marius Penin la Medaille commémorative de l'établissement du Système Métrique et de son usage exclusif." A specimen of the medal is in the Streeter Collection.

After the Franco-Prussian War, a kind of agreement was signed in Paris in 1875 at the conclusion of the first international convention of the meter

Fig. 1. Medal commemorating acceptance of the metric-decimal system in France, issued in 1840.

A

Fig. 2. Medal commemorating the International Commission of the Meter.

B

which, in fact, had begun before the War of 1870 and was resumed in 1872. This agreement under the name "Convention du Mètre" was signed by the participants of this long-lasting conference on May 20, 1875, and published in the same year in Paris. Another beautiful bronze medal, made by J. C. Chaplain in 1874 (diameter: 100 mm), was then dedicated by France to the scientists who had helped to bring this remarkable endeavor to a successful end. Still under the influence of the horrors of war, the obverse of this medal bears as dedication the words: POPULORUM CONCORDIAE SACRUM (Fig. 2).

Tables 1 and 2, giving denomination and comparison of the weights of the new metric standard with the standards in Canada, were published in 1873 in London in the seventh annual report of the Warden of the Standards.

TABLE 1. Metric Weights in Canadian Standard Equivalents

Metric Denominations and Values		Equivalent Expressed in Terms of the Standard of Canada	
Name	Grams	In Pounds Avoirdupois and Decimal Parts of a Pound	In Grains and Decimal Parts of a Grain Troy
Millier	1,000,000	2,204.62125	
Quintal	100,000	220.46212	
Myriagram	10,000	22.046212	
Kilogram	1,000	2.204621	
Hectogram	100	0.220462	
Decagram	10	0.022046	
Gram	1	0.002204	15.4323487
Decigram	1/10	0.0002204	1.5432348
Centigram	1/100	0.0000220	0.1543234
Milligram	1/1,000	0.0000022	0.0154323

TABLE 2. Denomination of Metric–Decimal Weights*

Grams	Greek or Latin Name and Value		Denomination of Weight
10,000	Myria	= 10,000	Myriagram
1,000	Chilia (kilo)	= 1,000	Kilogram
100	Hekaton (Hecto)	= 100	Hectogram
10	Deca	= 10	Decagram
1/10	Decem	= 10	Decigram
1/100	Centum	= 100	Centigram
1/1,000	Mille	= 1,000	Milligram

* The final nomenclature of the metric weights puts the gram in the central position. Its multiples are designated by Greek prefixes, its fractions by Latin prefixes. This combination leads to the denominations shown in Table 1. To these should be added the multiples of the kilogram (1,000 gm): 10 kg = 1 myriagram; 100 kg = 1 quintal; 1,000 kg = 1 metric ton or millier.

4. SCALES

It seems more than probable that weighing began before the invention of scales. If his physiological functions were normal, early man, lifting an object in each hand at the same time, must have had some idea of their relative weights. Egyptian monuments and papyri depict persons carrying shoulder yokes with containers attached. This kind of activity would have made men conscious of balancing weights and must have taught them not only all the physical principles of the balance but, even before the discovery of the laws of the lever by Archimedes, would have taught them that changing the pivotal point of the yoke would bring unequal loads into equilibrium. The see-saw of children has also been regarded as the real origin of the balance (Guichard, 1937; Sökeland, 1900). Thus practical coincidental experience apparently guided man to the invention of the indispensable instrument of weighing— the scales in their different forms—which will be described here.*

Every kind of scale serves basically the same purpose, which is to determine the weight of certain objects as exactly as possible. The objects vary greatly in size and mass, and the scales used must vary accordingly. The basic principles of construction are not many, and they distinguish the several types of scales:

> Balances (scales with beam arms of equal length)
> Steelyards (scales with beam arms of unequal length)
> Bismar (also with unequal beam arms)
> Spring scales
> Platform scales (decimal scales)
> Indicator scales (pendulum balances)

There will be no particular description of the modern precision scales of today because their basic principles are contained in the types mentioned above. Electrical scales and electronic microscales, with their admirable precision and technical perfection, are also omitted. An object is put onto the scale and everything else is done by the machine, including presentation of a ticket on which the exact weight of the object is neatly printed. These scales no longer need weights except for an occasional adjustment, and it is therefore doubtful that they can properly claim a place among the instruments discussed here as scales.

---

* For the names of scales in various ancient languages, see Machabey, pp. 15–22.

THE BALANCE*

The balance is a scale with two pans (in old English books also called scales) suspended from the ends of a beam. One pan is used for the goods to be weighed, the other for the weights. Equilibrium of the loads is indicated by an exactly horizontal position of the beam.

The Latin expression for this type of scale is *bilanx* (from *lanx*: dish, pan). Chart 1 shows the basic parts of a balance and their denominations as used in the following sections.

The first balances were unquestionably of this type: a pan or basket was attached to each end of a stick, one to hold the weights, the other for the matter to be weighed, and the stick was held on the hand or balanced on one finger and the equilibrium judged approximately. Thus the first beams were not suspended to swing freely but were supported from below. This is proved by Figure 3, one of the oldest illustrations of a balance known, on a Hittite

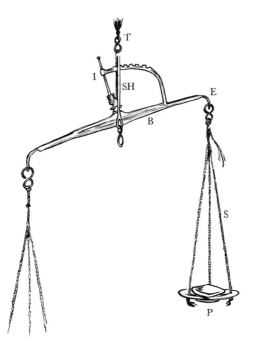

Chart 1. Parts of a balance-type scale. Only the right pan is indicated. B = beam; E = beam end; I = indicator; Sh = shears; T = tassel; S = strings; P = pans.

---

* Latin: *trutina, libra, bilanx*. German: *Balkenwaage*. French: *balance à bras égaux, balance de magasin*; if the beam is shorter than 65 cm: *balance de comptoir*.

27

relief found in Karkemish, which is now in the Louvre in Paris. The primitive scale is held from below; the little ornament on the beam above the hand may indicate the best place for the supporting hand or finger. A similar scale from the Roman period shows a deity with a balance, the beam being supported by the hand from below (Ibel, 1908, Fig. 15).

This kind of balance, despite its inexactness, was used for millennia according to Chambers, who, in the first edition of his famous *Cyclopaedia*, describes one such under the name of *auncelweight** or *handselweight* (literally as "consisting of scales hanging on hooks fastened at each end of a Beam or Staff which a man lifts up on his Hand or Forefinger and so discovers the equality or difference between the weight and the thing weighed"). Chambers also stated that such balances, which made cheating too easy, were prohibited by several statutes. He mentioned that auncelweight was still used in his day (1728) in some parts of England to signify meat sold "by poising in the hand without putting it into the scales." In Cologne, however, a large center of commerce from medieval times, the city government by 1348 had already prohibited the sale of meat without weighing it. This caused such a vehement revolt among the butchers that the City Council temporarily suspended the butchers' guild. The fishmongers of Cologne were ordered in December 1482 to sell salted fish exclusively according to weight, as had been the custom for some time with fresh fish. According to Chambers' statement, butchers in parts of England as late as the early eighteenth century still were not compelled by law to sell meat by weight.

Figure 4 shows the very different types and sizes of balances used from ancient times. There were small balances to be held in the hand, as in the Hittite relief. Ducros (1908) describes an ancient Egyptian balance which had a beam only 138 mm long. Small balances have been found in little pouches in Merovingian graves (Werner, 1954) and also in those of the Vikings (Lundquist, 1956; Kisch, 1959). They were probably used to weigh coins, gold, or jewels. There were also man-sized balances, suspended from special stands, with very long arms, probably to increase their sensitivity. Their appearance and use are clearly demonstrated in old Egyptian pictures, especially in the many surviving books of the dead (Fig. 4). These books usually showed the psychostasy (the weighing of the good and bad deeds of the deceased, both indicated by symbols). The big balances shown for this purpose probably represent, after the Hittite (Fig. 3) and the Assyrian sculpture (Fig. 5), the oldest pictures of this instrument and its use. Such books of the dead date from the second millennium B.C.

---

* In parts of Germany the name *Uenzel* was still used in the early twentieth century for the bismar (Sökeland, 1900; see also pp. 26, 56).

Fig. 3. Hittite relief. This type of scale was supported from below; note the basket form of the pans, supported by two strings. The meaning of the object in the left hand is not clear; it could be a container for weights.

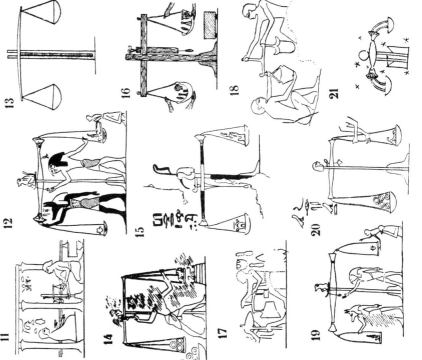

Fig. 4 Various Egyptian balances, as depicted in books of the dead.

Fig. 5. Assyrian relief on a stele from the reign of Ashurnasirpal II, 884–859 B.C.

The fact that Babylonian and Assyrian weights have been frequently found and with inscriptions that identify them as weights (see Figs. 74–76) is proof that scales of some kind existed in the fourth millennium B.C. and possibly even earlier. Of a later date but of great interest is the illustration of a scale and act of weighing from a stele from the reign of King Ashurnasirpal II, King of Assyria (884–859 B.C.), found at Nimrud by Rassam in 1853 (Fig. 5).*

All parts of the balance (shown in Chart 1) have undergone various transformations during the ages to improve their efficiency. The requirements for a good working balance were set forth by Chisholm in 1877 in his book on the science of weighing and measuring (p. 135), and are still valid today:

> For the justness of an equal armed balance it is requisite: (1) That the two points of suspension of the pans from the beam be exactly in the same line as the centre of motion or the fulcrum on which the beam turns when set in motion. The line joining these three points is the axis of the beam. (2) That these two points be exactly equidistant from the centre of motion. (3) That there be as little friction as possible at the centre of motion and the points of suspension. (4) That the centre of gravity of the beam be placed a little below the centre of motion.

The Beam†

It is thought that most scale beams were originally of wood, and this may be the reason that scarcely any early Egyptian and pre-Egyptian examples have survived. Although wood was usually the very early choice, stone or bone was used on rare occasions. The collection of Flinders Petrie at University College, London, contains a very early beam from an Egyptian balance (ca. 3000 B.C.) made of brown limestone (Glanville, 1937). Even in 1773 Jaubert says (p. 192) that the beam of a balance in France was of copper, iron, or wood; this was also true in Cologne in the eighteenth century (Kisch, 1960b), and Rome had a preference for brass, as can be seen from the many Roman balances found in Pompeii. Wooden beams were still used in Europe in the nineteenth century, as shown in various specimens in the Streeter Collection. In Cologne there are examples of large commercial balances of this type dating from the eighteenth century (Fig. 47). There is also confirmation

---

* I owe this picture and information to the kindness of Dr. Sollberger of the British Museum, where the stele is preserved. The picture is reproduced through the courtesy of the Trustees of the Museum.

† Latin: *jugum*; its branches, *jugi brachia*. Medieval Latin: *virgula, regula*. Old English: branches. French: *fléau, flayau*, or *flayan*; also *trabeau, traversin, basque*. German: *Balken, Waagebalken*. Italian: *giogo*.

of this in repeated decisions of the Cologne government concerning the competence of the craftsmen who were to be permitted to make and adjust balances with wooden beams with or without iron mountings (Kisch, 1960b). In Lyons the use of wooden balances was forbidden in 1781 (Machabey, 1962, p. 320).

Iron beams were used for the large public scales sponsored by the government and located in the marketplaces in various European cities. For the fine balances of pharmacists, jewelers, or bankers the beam was usually made of brass or polished steel to prevent rust. For the same reason the beam and the pans of small balances (French: *trébuchet*, also *biquet*) were made of silver, ivory, or similar precious material. The beams of the pre-Columbian Peruvian scales were made of bone and those of the so-called Chinese opium scales of ivory.

The beam was originally cylindrical (see Fig. 3) or flat, later tapering at both ends. Pictures of Egyptian scales often show a widening of the arms in the form of a trumpet or lotus flower, as does the beam of the Assyrian scale (Fig. 5). The beam itself was frequently colored and ornamented. (See the very instructive paper of Ducros; it has pictures of forty-nine Egyptian balances, from which Figure 4 is taken.)

Fig. 6. Weighing scene on the cup of Arcésilas, King of Cyrène, sixth century B.C. Note the great similarity to Egyptian weighing scenes in Figure 4.

33

The balances used in Egyptian commerce were of the same type as the symbolic ones found in the books of the dead, as can be seen in secular pictures from the same era in which certain goods (probably rings or cylinders of certain metals) are weighed with weights in the form of different animals (cattle, hippopotamus, etc.; see Figure 4).

The beam of a scale pictured on a Greek cup (in the Cabinet des Medailles et des Antiques at the Bibliothèque Nationale in Paris), designated the "Cup of Arcésilas, King of Cyrène," from the sixth century B.C. (Fig. 6), is similar to the beam of the Egyptian balance. Not only the balance but the entire scene of weighing seems to have been strongly influenced by the Egyptians.

A cord slung around the middle of the beam and hung on a stand or held by the hand was probably used originally as a pivoting point in these balances. This method permitted a changing pivotal point or fulcrum, which survived in spite of its lack of exactness. It is used today in the type of scale called the bismar or desmar (pp. 56 ff.). However, in ancient Egypt there were already balances with fixed fulcra. The beam was pierced in the middle, a cord run through, and the cord held in the hand or suspended from a hook.

This type of primitive scale was also invented by other peoples, independent of Egyptian influence—for instance by the Peruvians, as can be seen in Figure 7. Some Peruvian scales of this type are in the Streeter Collection. Kelemen in his book on medieval American art (1943, Plate 299) shows a beam very similar to those in Figure 7.

To produce a good balance, the position of the pivoting point in the beam is an important problem, as is the attachment of the beam to a stand. The suspension of the Greek scale on the Cyrene cup was from the brass ring in the middle of the balance beam. This ring held another ring or a cord to suspend the beam. A final form of fulcrum* was a suspension of the beam with an indicator from attached shears. The shears were also called *trutina* in Latin and in Italian. The indicator or pointer clearly shows the equilibrium or lack of it. It took millennia before this type of fulcrum, commonly used in Roman times, was supplanted by supporting the beam and the pans by knife-edges of metal (preferably steel), of agate, or recently of sapphire (Mettler *News*, 1, p. 7 ff.).

Egyptian balances had some kind of pointer, too, although the stylistic drawings depicting them do not always give an exact idea of how they used the plummet (a heavy object hanging from the stand of the balance to check the position of the beam and its indicator). In most Egyptian pictures of weighing (see Fig. 4) the man who apparently is the weighmaster points to

---

* Fulcrum in German: *Axe*. Italian: *asse di rotazione* (Aristotle's *sparthum*; Lazzarini, p. 221). French: *axe, chef*.

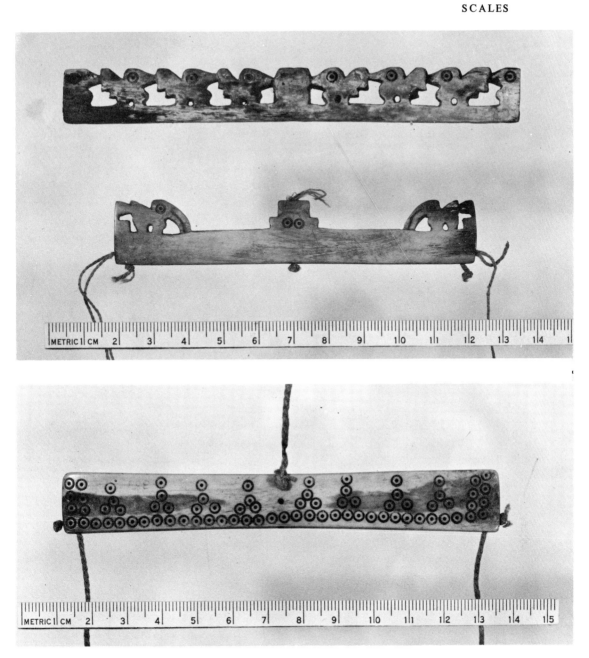

Fig. 7. Peruvian balance beams from the pre-Columbian period.

the plummets as if to indicate the correctness of the weighing by calling attention to them. The same scene appears on the Cyrène cup. The plummet has not been attached to the balance but to its stand. An indicator in the form of a kind of ruler which points downward has been fixed at a right angle to the beam. The big Assyrian balance (Fig. 5) shows no plummet and no indicator but an elaborate, artistically adorned stand. The Roman type of indicator (a pointer going up from the beam at an angle of 90°) remains the standard for many modern balances used in marketplaces.

The indicator or pointer of a balance is called, like the shears, in Latin: *trutina*; also *examen* or *lingula*. In German it is called *Zunge* or *Zünglein*. In Italian it has various names (*Istruzione*, 1750): *perpendicolo, ago, aguiccietta, guidice, guisticia*; in French the word is *aiguille*. In modern analytical balances the indicator again points downward as in the Egyptian scales, but it shows its deviations on a millimetric scale attached to the stand.

In the Roman and Byzantine periods the fulcrum was often not in the beam itself but above the center of the beam in the indicator, in the form of a hole through which an axis was fitted as part of the shears (Figs. 8, 9).

It appears that the beam without a special pointer (Fig. 7) is the more primitive and therefore older instrument, but excavations at Pompeii prove that balance beams both with and without shears and pointers were already in use before A.D. 79. One often finds on one arm of the beam of Roman balances a regularly spaced row of dots and lines indicating that these balances, like modern analytical scales used by chemists, were used with a rider-counterpoise in addition to the weights on the pans. This will be discussed along with steelyards.

The attachment of the pans to the beam was provided by three or four cords or chains. The manner in which they were supported and connected with the beam, a point of high importance for the correctness of weighing, was basically the same as discussed for the fulcrum of the beam. It went through all the stages of improvement from the primitive loop of cord to the agate knife-edge support. In modern times the suspension of the pans from the beam has been changed. In the platform equal-arm balances the platforms and pans were put above the beam, the ends of which had provision for supporting the pans. Of some interest and help in determining the origin of a beam is the form of the beam ends to which the pan strings have been attached. In the earliest known (Egyptian) scales the strings were attached directly to the beam and to the pans. The Peruvian scales in the Streeter Collection (Fig. 7) and those in the Miguel Mujica Gallo Collection* show the same primitive attachment.

---

* The collection was exhibited in Cologne in 1959 and a catalogue of it has been published (*Schätze aus Peru*, 1959).

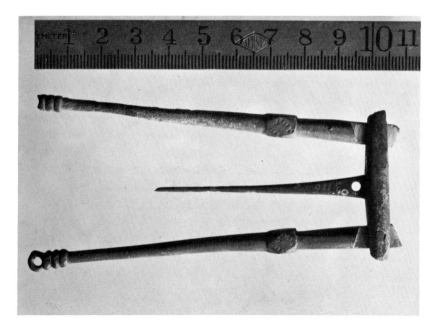

Fig. 8. Byzantine folding scale. The fulcrum (axis) is not in the beam but above it in the indicator.

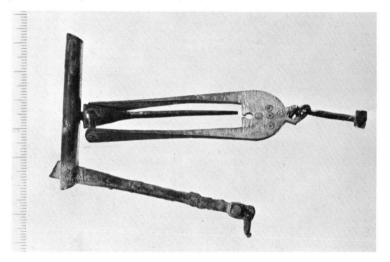

Fig. 9. Folding scale from the Roman period, similar to the scale in Figure 8.

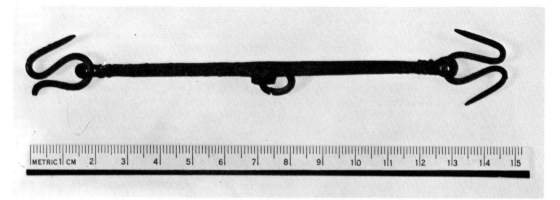

Fig. 10. Roman hand scale, probably excavated in Egypt. Note the double hooks to support the pans, common on Roman balances found in Pompeii.

Later, as a rule, a metal ring or hook was attached to each end of the beam to hold the strings or chains that supported the pans. The hook was often in the form of a question mark or a figure eight, but in Roman scales it was frequently an interesting kind of double hook, which was already in use before A.D. 79 and is commonly found on scales excavated in Pompeii; it is still seen in the late imperial and Byzantine periods, as is apparent in Figure 10. A few typical beam ends used in different countries at different times, especially for the little gold scales, are shown in Chart 2.

Some special types of balance beams should also be mentioned. One is the collapsible beam, which has a joint in each arm equidistant from the fulcrum, permitting the beam to be folded when not in use so that it can be carried comfortably in a small container in the pocket. This type of balance was used in Roman and Byzantine times and appears in Egyptian excavations

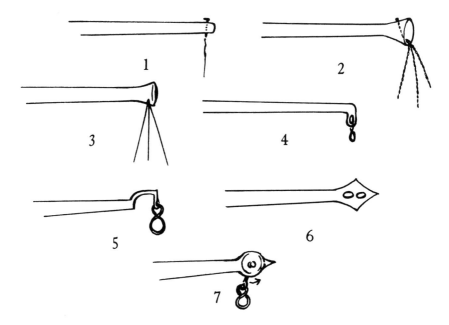

Chart 2. Various types of beam ends for gold scales. (1) Most primitive form used in early Egyptian and Peruvian balances. Attached to it are baskets (Peru) instead of pans or a hook (Egypt). (2) A more elaborate similar type with three strings. (3) Trumpet form of beam end, common in Cologne in the sixteenth and early seventeenth centuries. (4) Spatula form of beam end, common in Cologne and elsewhere in Germany in the seventeenth century. (5) Gallows or swan-neck form of beam end, used often in Europe in the eighteenth century. (6) Rhomboid form with two holes through which a hook was suspended, carrying the strings of the pan; this type was very common in French scales of the eighteenth century. (7) Box end of the scale used in the eighteenth and early nineteenth centuries in England and in Germany, Austria, and other European countries.

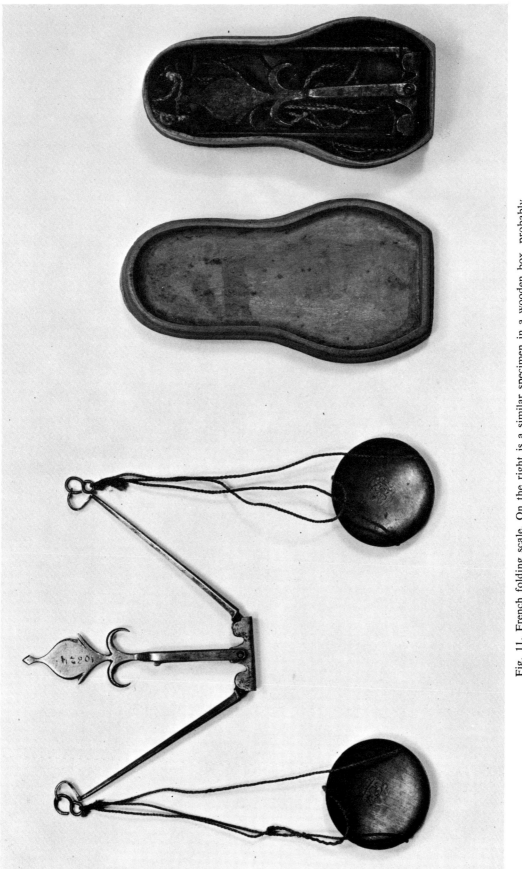

Fig. 11. French folding scale. On the right is a similar specimen in a wooden box, probably dating from the end of the seventeenth century.

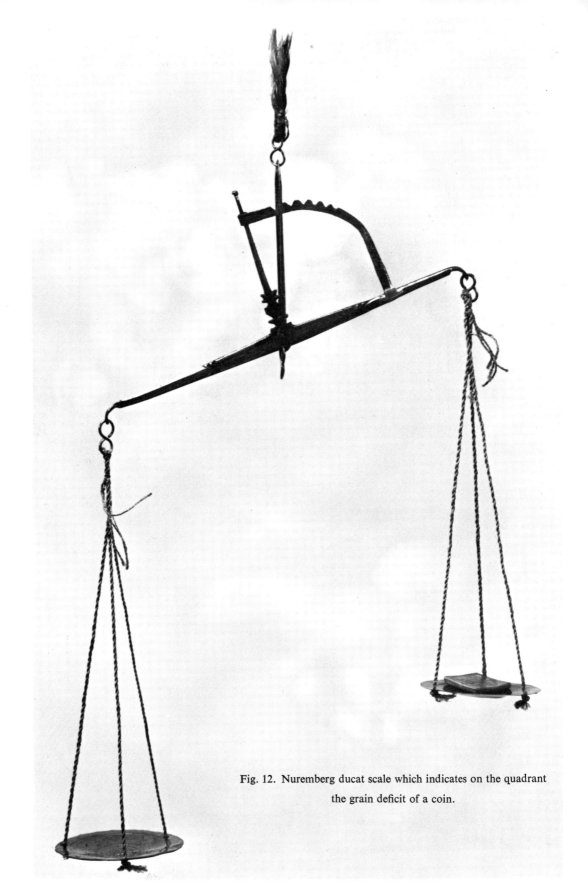

Fig. 12. Nuremberg ducat scale which indicates on the quadrant
the grain deficit of a coin.

from this period. Figure 8 shows such a scale and Figure 9 shows part of a similar scale, which was found among other Roman relics excavated in Mainz, the old Roman Castell Moguntia. The same type of collapsible balance was found in Viking graves and can be seen in the Royal Mint display of the National Museum of Stockholm. The Streeter Collection has a similar French balance, probably from the sixteenth or seventeenth century; two very similar specimens of French collapsible balances from the seventeenth century with the original pyriform container intact are in the collection of the Wellcome Historical Medical Museum in London (Fig. 11). In both these scales the maker's mastermark is impressed in the pans. This was apparently a very practical type of balance and remained in use for a long time.

Another practical beam modification was very popular in the eighteenth century and is pictured in Figure 12. One arm of the beam is connected with a subdivided arch which in turn is connected with the pointer. If the balance is not in equilibrium, the shear works as an indicator, pointing to a certain part of the arch where the amount of deviation can be immediately determined. These combined balance and indicator scales were much used for the so-called ducat scales made mainly in Nuremberg during the eighteenth century and will be discussed in detail (p. 67). They were usually used for weighing Hungarian gold ducats or other gold coins. One pan was just one ducat heavier than the other so that a merchant could quickly determine whether a coin was of honest weight. If there was equilibrium with a coin in the lighter pan, the pointer stood just within the shear. If the ducat was underweight from wear and tear or from clipping, the pointer immediately indicated on the arch (see Fig. 12), how many grains of gold were missing from the legally prescribed weight. These ducat scales were probably the prototype of all the various indicator scales that are used today in innumerable varieties and have the advantage that they require no weights.

Other changes in the form of the beam were suggested later, and some of these modifications are in use today to make the balance more sensitive and more reliable, especially for analytical purposes in chemistry. One of the first of these was an analytical balance of high precision proposed by Sir George Shuckburgh Evelyn in a meeting of the Royal Society in London in 1798 (Fig. 13). The modification in the shape of the beam in modern analytical scales is more efficient than the awkward Evelyn model. Figure 14 shows a complex modern analytical balance made by Sartorius in Göttingen.

The Strings

According to Jaubert (1773, p. 193) the strings or chains from which each pan of a balance is suspended are supposed to measure in length twice the diameter of the pan. The strings have numbered either three or four since ancient times. Only the oldest pictures of Egyptian balances, from about 3000 B.C., show a single string ending in a hook at each end of the beam, from which the load and the weight were hung (Glanville, 1934; see Fig. 4). A copper pan found in Megiddo shows three holes for suspension (Stern, 1963, p. 543). In the Roman period, three or four strings were used. Later, as a rule, the pans in small balances were attached with three strings in European countries and with four in the Near and Far East. The four holes in all extant pans of Egyptian balances prove that they were suspended from four strings. Ducros (1908, p. 37) rightly emphasizes that old Egyptian pictures showing pans suspended from only two strings are merely symbolic representations. For any practical purpose the suspension of a circular pan from the beam of a balance by two strings is not feasible. It is possible, of course, that metal rods were used on some ancient balances, as they are used today.

If a basket is substituted for the pans, as in the Peruvian and some early Egyptian scales, or even in the Assyrian (Fig. 5), suspension from the beam end by two strings is of course possible.

Four chains were used as a rule in Europe to suspend the pans of large balances for heavy loads; this is portrayed in a famous relief of 1497 (Fig. 15), which shows all the details of a large balance operated by the city of Nuremberg. The merchant is about to take from his purse the fee owed to the weighmaster, who is in the center watching the weighing procedure, while the young apprentice at the left puts the heavy weights onto the quadrangular pan of the scale. Metal chains were used in Rome for balances even of small size, as specimens from Pompeii show, and Figures 16 and 24 depict chains of different types.

If cords rather than chains were used on sensitive scales, silk was the preferred material, probably to prevent changes in weight owing to hygroscopic attraction of water from the air or by any fraudulent manipulations. The government of Cologne prescribed silk strings for gold scales (Kisch, 1960b).

The silk strings of fine balances in Europe were almost always green. The reason for this preference can only be suspected. Ercker, in a famous book on mineralogy (1574), advised the assayer to relieve the strain on his eyes by putting his scale into a green cabinet because this color was supposed to be an especially relaxing one for the eyes. Perhaps for the same reason the balance makers in Europe used green for the balance strings. Yellow was used less often, and in the Near and Far East red was apparently preferred.

In modern times in pharmaceutical and analytical balances strings or chains are replaced as a rule by stiff metallic rods suspended from the beam by agate supports in the form of knife-edges (Fig. 14).

Fig. 13. Precision balance proposed in 1798 by Evelyn and published in *Philosophical Transactions of the Royal Society.*

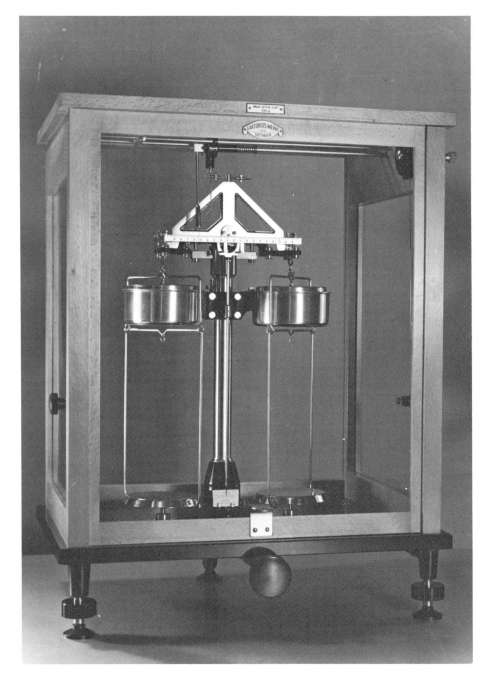

Fig. 14. Modern precision analytical scale.

Fig. 15. The weighing process in Nuremberg in 1497. This relief was above the gate of one of the public weighing offices.

Fig. 16. Steelyards from Pompeii with their typical counterpoises.

*The Balance Pans**

After the beam, the pans of the balance next command the attention of the balance maker because they too are responsible for exactitude in weighing. The pans as well as the beam probably were originally made of wood, although in some ancient Egyptian pictures, baskets of different materials are shown. Wooden pans are still used, especially for weighing heavy material when high exactness is not essential, and examples of such scales from European countries are preserved in the Streeter Collection. More stable and less hygroscopic, of course, are metal pans. Relics from the Roman period show that iron as well as brass was used. Silver pans are occasionally found in later times; in the collection in the Louvre in Paris there is a lavish balance, probably from the seventeenth century, that has beam and pans of gilded silver.

For pharmaceutical purposes, materials that could be easily cleaned have usually been preferred. In England in the eighteenth century, pans were often made of glass, sometimes of ivory, or (rarely) porcelain. A German pharmaceutical scale of the nineteenth century, which has china pans, each bearing the inscription *Gift* (poison), has recently been acquired by the Smithsonian Institution in Washington. Pans infrequently were made also from mother-of-pearl. In the nineteenth century and up to the present, simple laboratory scales are often equipped with pans made from horn, which is also easily cleaned.

Balances with double pans (one pair attached to the beam by the strings, the other resting on the first ones and easily removable) were used during the eighteenth, nineteenth, and twentieth centuries. This double-pan arrangement with two glass pans on each side was usually for small pharmaceutical scales to be held in the hand. Examples of such small balances from eighteenth-century England are in the splendid collection of the Wellcome Medical Museum in London. The Streeter Collection also has one set. Figure 17 shows a gold scale of this type from Austria. Of course it is necessary that each of the two fixed and each of the two removable pans should weigh exactly the same as its counterpart.

The exception to this rule is the balance used for the rapid weighing of coins, such as the Nuremberg ducat scales (Fig. 18). These scales were called and usually also inscribed: "Ein Wäglein ohne Gewicht" (a small balance without [removable] weights).

The oldest known examples of such balances are again from the Roman period. Lorenzo (1735) describes two Roman balances, kept in the Museo

---

* German: *Schalen*. French: *bassins, plateaux*. Italian: *coppe* or *piatelli* (the flat ones, *piatti piani*, the concave ones, *piatti a coppa*); in Old English pans were also called scales.

Archeologico in Florence, and Ibel (1908, p. 60) shows one of them as do Sheppard and Musham (1924; their Fig.4). The balance has a pan on one side and on the other, suspended by a chain from the end of the beam, the head of Juno Moneta. This head was heavier, by exactly the weight of a particular coin, than the pan on the other end of the beam. According to Lorenzo, in a certain case this was a quadruple denar of gold. This scale in the early literature was called *trutina momentana* (instantaneous scale), and because it was used for weighing money it was also called *moneta* for short. These monetae were already in use in the time of the Emperor Honorius (384–423). The Juno Moneta face on these scales was a kind of stereotype. The Streeter Collection possesses one such weight (without the scale) which looks exactly like the ones depicted by Ibel and by Sheppard and Musham (Fig. 19).

The Roman trutina momentana was also used later in medieval Germany and called *Seiger*. From the end of the thirteenth century the city law of Freiburg specified that this kind of balance be used for detecting overweight silver coins. Unlawful possession of a Seiger was punished by severing the hand of the culprit (Hoops, 1918/19).

Arresting the Balance

When a balance is not in use it is desirable to immobilize the beam to avoid unnecessary wear on the points of suspension (knife-edges, fulcra). The most primitive way to do this, of course, was to put the balance on a flat surface. A further refinement was to hang the scale high enough so that the pans did not touch the ground when it was being used but were grounded if not in use. The Talmud prescribes (Baba Bathra 89 a) the height of the pans of a balance above the floor during the process of weighing. This was different for different kinds of balances, apparently dependent on their size. It was three palms for the balances of wool dealers and glaziers; the size of the beam and the length of the cords were in this case nine palms each. The balance of the merchant and of the housefather (i.e. for home use) was supposed to have a beam only six palms long, and in equilibrium the pans had to be only one palm distant from the floor. Apparently these customs were derived from the practices of the Roman market.

A typical device to lift the balance pans from their support before the balance could be used is of special interest and is found on the pharmaceutical scales of Nuremberg in the seventeenth, eighteenth, and early nineteenth centuries. It is shown in Figure 20. A wooden box contains all the weights and also houses the balance and a column that can be fixed upright on the box. Atop the column is a two-arm lever from one end of which the balance is suspended. A string is attached to the end of the other arm; when the string

48

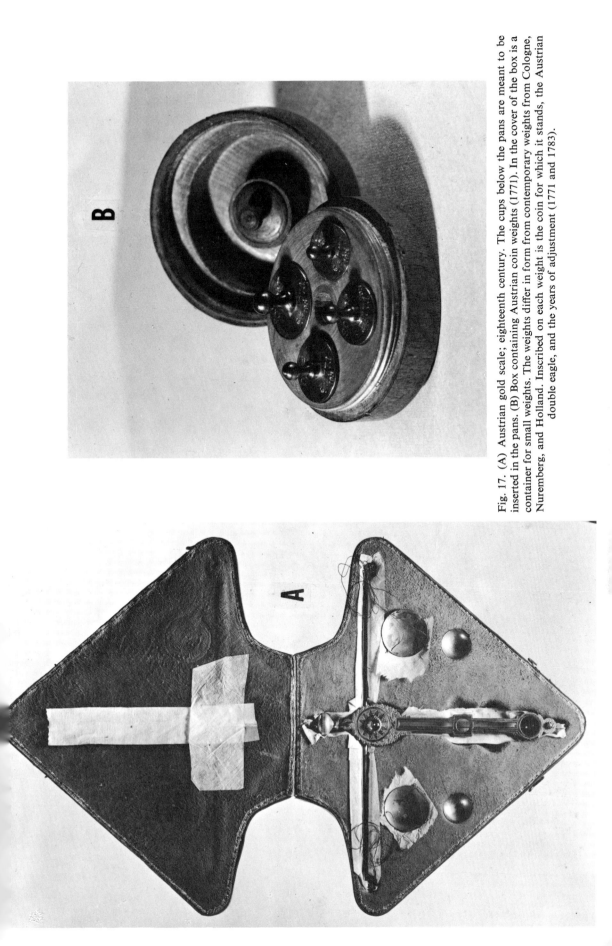

Fig. 17. (A) Austrian gold scale; eighteenth century. The cups below the pans are meant to be inserted in the pans. (B) Box containing Austrian coin weights (1771). In the cover of the box is a container for small weights. The weights differ in form from contemporary weights from Cologne, Nuremberg, and Holland. Inscribed on each weight is the coin for which it stands, the Austrian double eagle, and the years of adjustment (1771 and 1783).

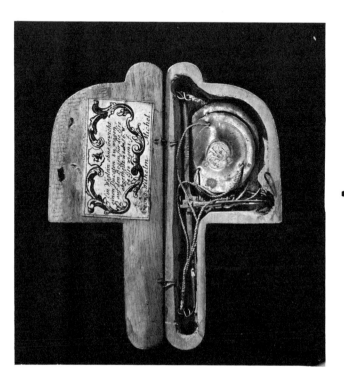

Fig. 18. (A) Ducat scale made by Wilhelm Michel in Nuremberg in the eighteenth century. (B) The two pans of a Nuremberg ducat scale; in the bottom of the one at the left is a symbol of the Hungarian ducat (HD), in the other the mastersign of the balance maker (two crossed arrows and initials IS).

is pulled, the arm with the balance rises and lifts the balance so it can be used. If the tension on the string is lessened, the arm bearing the balance goes down, the pans of the balance touch the box, and it is thus arrested. The string to free or arrest the balance was not held, because the operator needed both hands for the weighing process. A weight, usually shaped like a lion, was attached to the end of the string and was put in place as soon as the column was erected. Its position was adjusted for arrest or release of the balance as required. This way of arresting a balance is also shown in three scales of different sizes in Agricola's *Berckwerk Buch* (1580, p. 214).

Another simpler way of preventing wear when a balance is not in use is to leave the balance suspended and put one pan into the other pan. This simple action keeps the beam from swinging and was probably well known to the Greeks. Homer (*Iliad*, VIII) describes how Zeus used a scale to weigh the fate of Greece and Troy, separating the pans of his golden balance from each other and putting a lot into each pan. Homer's description indicates that the Greeks knew the two-arm balance (and possibly also balances made from gold) and that the pans of the balance, when not in use, were stacked, as is done in the laboratory today with the simple horn balances.

A third way to arrest a balance is to lift or lower the base on which it rests rather than the balance itself (Fig. 13). Many different modifications of this principle are used today.

Damping the Balance

Devices for damping a balance are mainly a modern problem. Damping in any mechanical system means to cause a diminution of the amplitude of its oscillation in order to bring it quickly to rest.

Fig. 19. Counterpoise of a *trutina momentana* from the Roman period, for weighing gold coins without individual weights.

51

For modern laboratory scales this problem is important to ensure rapid weighing without reducing exactness. Details of the damping devices for modern balances are very well described in the Mettler *News* (1, pp. 18–27), and are easily understandable by the layman. They are distinguished as: (1) inertial damping, (2) air damping, (3) liquid damping, and (4) eddy-current or magnetic damping.

Balances for Specific Purposes

Scale makers have always been faced with the problem of constructing balances to fit the different purposes for which they were ordered. Balances for gold or jewels naturally had to be of other dimensions than those for weighing heavy objects. The balances depicted in the Egyptian books of the dead are the size of a man and were supposed to weigh on one pan an ostrich feather (symbol of the goddess Maa), on the other the deeds of the deceased.

Figure 5 shows a large Assyrian balance of the ninth century B.C. suspended from an elaborate and artistically adorned stand. In the trumpet form of the beam ends and the shape of the beam it resembles ancient Egyptian scales (Fig. 4, nos. 12, 19, 20, 22). It is the height of a man and was used for weighing heavy goods. It may represent the weighing of tribute (bricks of some metal) because the two persons involved in the act of weighing look as though they might be of different nationalities—one with and one without a beard, different size, and different attire. In the middle of the beam, whose type of suspension is not clearly indicated, is a small ornament similar to that in the middle of the small beam of the Hittite scale (Fig. 3).

That large and small balances were used in ancient Rome to weigh different objects even within the same establishment is proven by the murals found in Pompeii showing putos working at the mint.

The Talmud (Baba Bathra 89 a/b), in its regulation of the use of weights and scales, mentions various scales for weighing such different substances as base metal, wool, and gold; it also alludes to scales used by glaziers. Whether they sold glassware according to its weight or bought their raw material in this way is not clear from the text.

Later and up to the present we find not only the size of the balance but also the shape and, often, the material of the pans adapted to the particular purpose of the balance. Candles, for instance, which were bought and sold according to weight up to the eighteenth century could not be laid in a bowl-shaped pan for fear of breakage and so we find special flat pans on balances used particularly for candles. The balances of money changers also had characteristic pans, and will be discussed in a later section.

52

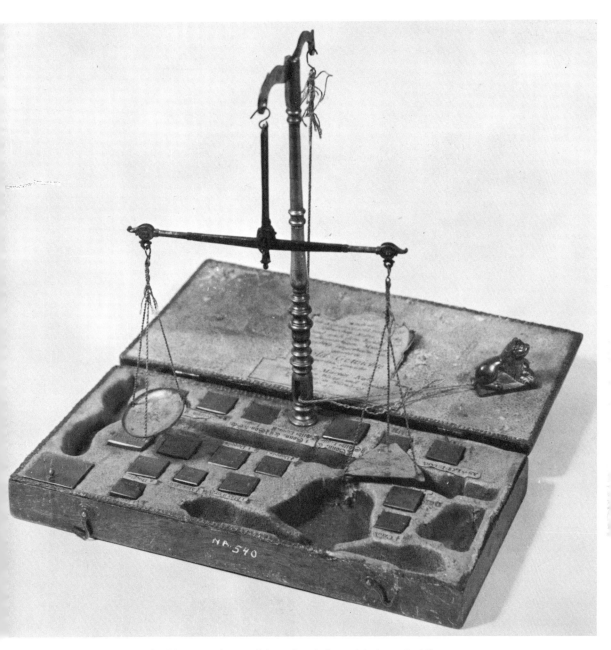

Fig. 20. Nuremberg weight and scale box with the typical lion.

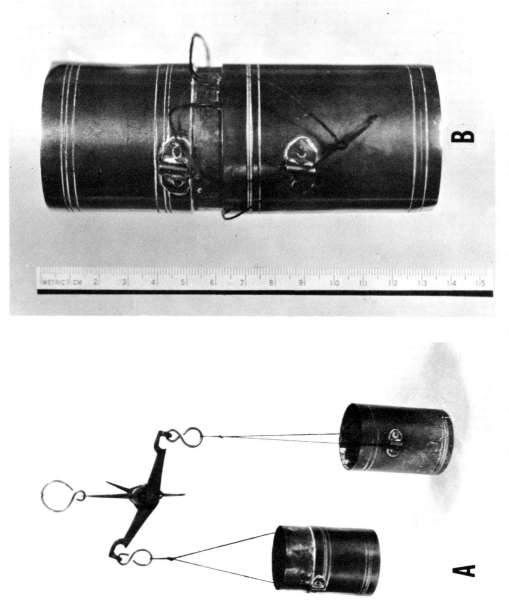

Fig. 21. (A) Dutch grain scale (*Kornwaage*). (B) The same scale; the beam and the weights are in the closed box.

A very unusual type of balance pan has been used up to modern times in Holland for the purpose of judging the quality of a certain measure of grain by its weight. These balances, usually small, had cylindrical pans (Fig. 21) that could be fitted together to make a container for the small beam and the weights, which were of typical mushroom form. This kind of scale could be carried in the owner's pocket and was apparently used only in Holland. It was called *balance d'Amsterdam* or *balance hollandaise* and was popular in such towns as Amsterdam, Antwerp, Hamburg, and Riga (Doursther, 1840). Leupold (1726, p. 60) shows a scale like this (Kornwaage), with one cylindrical pan which is a steelyard, not a balance.

In a book published in 1698 by one Christoff Weigel in Regensburg, a chapter is dedicated to the weight makers and the scale makers. Weigel describes the different types of scales used in his day in Germany. They are the following:

1. Common and fine balances, the latter called Cologne scales, used by merchants to weigh heavy goods in hundredweight amounts.
2. Balances for grocers and for spicers (*Krämer- und Gewürzwaagen*) with pans of brass and copper.
3. Candle balances, which had one round, concave pan and one flat one.
4. Good Cologne silk balances; the pans were also flat but with a slightly elevated brim.
5. Gold balances.
6. Steelyards.
7. Pearl and diamond balances.
8. Balances for assayers and miners.
9. *Notiometra*. An adequate explanation for this word could not be found. Weigel says that these balances had a small quadrant at the beam and were used to investigate the weight of gases. Zedler's *Universallexicon* (1747, vol. 52) mentions a Jesuit by the name of Lemu as the inventor of this type of balance.
10. Finally, as a "new" invention, Weigel mentions the construction of spring balances.

In the nearly three hundred years since Weigel, many other special balances have been added to this group, including the fine balances used in analytical chemistry, torsion balances, and the electrical balances.

Details of the construction of modern balances for industrial and scientific purposes as used in our time can be found in Brauer's instructive book, with 246 illustrations, published in 1909 in English by the Society of Inspectors of Weights and Measures in London and in the recent book by the Russian, Karpin (1960). The books by Owen (1922), Testut (1949), and Machabey

55

(Chap. 6, 1949) also contain valuable information of this kind. Many more elaborate treatises exist today on the different types of modern balances, their construction, production, and theory. Every handbook of technical engineering contains a chapter on balances. Testut issued a good bibliography up to 1946 and so did Machabey (1949); see also among many others Gotz, 1931; Karpin, 1960; Koehnle, 1910; Labielle, 1932; Sartorius, 1961.

Today the problem of weighing very heavy or bulky material is fairly well solved (see the section on platform scales). Scale makers now focus their interest on microscales, which will weigh reliably the smallest imaginable masses. A description of all the principles involved in these (electricity, piezometry, electronics) is beyond the scope of this monograph (see Sartorius, 1961; Mettler *News*).

BISMAR AND STEELYARD*

In an instructive two-volume work on physical instruments used for experimentation, Sigaud de la Fond (1775) starts the chapter on scales with the words: "The scales are of two types, the ordinary one and the Roman."

The bismar and the steelyard (called also *romana* or *statera*) are the two types of scales using the principle of a lever with unequal arms for the beam. This type, unlike the equal-arm balance, does not use weights placed on one pan to indicate the weight of the object lying on the other. All scales with unequal arms use a single counterpoise,† fitting only *one* specific scale.

There are two possible methods of weighing with an unequal-arm beam. The first is represented by the bismar, the second by the steelyard. The counterpoise of the bismar is fixed to one end of the beam and the axis of the beam can be quite simply shifted, for example, by a loose loop of cord supporting the beam, which can be moved along it, thus changing the ratio of length of the two arms of the beam. As soon as the beam is horizontal, indicating that equilibrium has been reached, the correct weight can be read on the beam from the position of the loop of cord (the fulcrum). The scale using this principle of immovable counterpoise and movable fulcrum in ancient times was called *desem* or *desemer*, also *besem*, *besemer*, or *bismar*; in Sweden it was called *besman*. The English use the word *besom*; the old German name used in the Altmark was *Uenzel* (Sökeland, 1900). It is mentioned in the seventeenth century by Simienowicz (1676, p. 32). Figure 22, taken from Olaus Magnus'

---

* German: *Schnellwaage*. French: *romain, trona, troneau, trosniel, traineau, croc, crochet*.

† German: *Laufgewicht*. French: *courseur, masse de compensation, contrepoid*. Latin: *equipondium*. Italian: *contrapeso mobile*.

Fig. 22. This is probably the earliest published picture of a bismar (from Olaus Magnus, 1555). Observe the large balance at the right and the weights of different size.

Fig. 23. Bismar of primitive form, very similar to that in Figure 22, stemming from Orsa in Sweden; probably seventeenth century.

history of the Scandinavian countries, is probably the first picture of a functioning bismar to have been published in a printed book (1555). The picture also shows a large balance and weights of different size. The time and place of origin of the bismar is unknown; in Sweden its beam was usually made of wood.

Sam Owen Jansson, curator of the Nordiska Museet in Stockholm, has called our attention to some classical literature which proves that the bismar was well described in the pseudo-Aristotelian work *Mechanica 20* (853 b 25) and was used in Grecian times (Jüthner, p. 201; Jansson, pp. 13–14) to weigh meat in the market.* It was later improved, and rare specimens from the Roman period have been preserved (Fig. 23). Figures 9 and 10 in Nowotny are pictures of peculiar Roman scales which were recognized by Jüthner as bismars.

Since the *Mechanica* contains no reference to the steelyard, it may reasonably be concluded that the bismar was used in Greece in Aristotelian times, but the Romans apparently substituted the steelyard for this inexact instrument (see Lazzarini, 1948, and Jansson's article "Besman" in his metrological dictionary *Måttordbok*, 1950). In Lunier's dictionary (1806, vol. 1, p. 133) he describes the bismar as "Peson danois ou suédois" and in use then in Denmark and Sweden. It had probably been used in Denmark and Britain from the time of the Vikings until it was made illegal in Britain by Henry II (1133–89) and again by Edward III (1327–77). Leupold describes and depicts in his book *Theatrum staticum* (1726, p. 57) a handy gold scale built on the principle of a bismar, as described by one Bardonneau in 1680. The bismar was also used in Russia up to modern times as is proved by a fine specimen from the nineteenth century, made of brass and bearing a Russian inscription, which is in the Streeter Collection. According to Sanders, the desmar was and still is in use not only in the Scandinavian countries, including Finland, but also in Burma, India, and the Malay Peninsula.

The steelyards, however, are more important than the bismar and are better known. According to Sökeland this type of scale came by its name in comparatively modern times because its first use in England was on the left bank of the Thames where the merchants of the German Hanse sold steel. They are based on the second principle of unequal-arm beams. Steelyards are levers with an immovable axis (fulcrum) which have a pan or hook or both attached to the shorter arm for holding the load and, suspended from the longer arm, a movable counterpoise. At the place where the counterpoise

---

* Dr. Jerry Stannard of Yale University was kind enough to check the literature concerning the Aristotelian *Mechanica*. The expression used for the bismar in the *Mechanica* is the plural αἱ φάλαγγες (*falanges*), and it seems to be the consensus of experts that the *Mechanica* was not written by Aristotle but was a product of the peripatetic school.

58

Fig. 24. Steelyards from Pompeii.

keeps the beam exactly horizontal, the weight of the object is indicated on an engraved or inlaid scale.

There is also an inversion of the steelyard. The axis of the beam and the weight are not movable, but the hook with the load can be moved along the longer arm of the beam; its position in equilibrium indicates the weight of the load. Such scales are supposed to have been used in ancient Egypt (Lazzarini, p. 229).

It is assumed that the Romans invented the steelyard because there is no proof of its earlier existence in either literature or art, but many spectacular specimens have been found in Pompeii, proving that the Romans used this type of scale 2,000 years ago (Figs. 16, 24).

This scale spread from Rome all over the world, and it is known everywhere as the Roman scale or *romana*. In Roman times it was called *statera*, or *trutina campana*, probably from Campania (Italy) where it is supposed to have been first used, according to St. Isidore of Seville (560–636). It was also called *trutina momentana* because it registered weight so quickly. This name was also used for another kind of scale (p. 48).

Although popular belief has connected the name romana with the Romans, there is some disagreement on this point. The counterpoises of ancient steelyards were occasionally in the form of a pomegranate, and it is true that the Arabian word for pomegranate is *rummana* (in Hebrew, *rimonim*). Leupold (p. 36), who accepts both etymologies of the name, reports as the source of the Arabic name the *Mechanica* (1671) of one Joh. Wallis. But neither on the page quoted by Leupold nor in the chapter on scales can such a statement by Wallis be found, and it seems very doubtful that the name romana should stem from an Arabic expression associated with one of the many different forms of the counterpoise rather than from the people who invented the scale and sent it all over the world. Leupold's quotation from Wallis (p. 109) reads in fact literally: "Statera quam (ob usum eius frequentum Romae) Romanam vocant" (The statera which (owing to its frequent use in Rome) is called Romana).

Figures 16 and 24 show a few typical specimens of steelyards found in Pompeii. Brass was used for the beam and the pans and usually for the chains. This does not mean that wood was not sometimes used for beams, but probably these would not have survived from Roman times. That wood was used elsewhere for beams is a fact, and in parts of Switzerland steelyards with wooden beams have been popular up to modern times. Very often at the end of these beams one finds a burnt-in sealer's mark (hallmark) bearing the arms of the canton where the scale was made. As with balances, the size of the steelyard depends on the average weight and size of the goods to be handled. Small steelyards, probably for weighing gold, coins, or jewels were usually made of

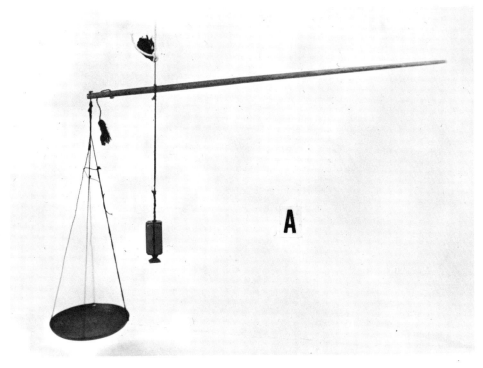

Fig. 25. (A) Chinese "opium" scale. The beam is of ivory, the weight is of typical form. (B) Violin-shaped container for the opium scale.

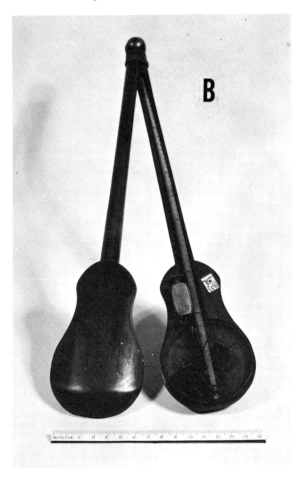

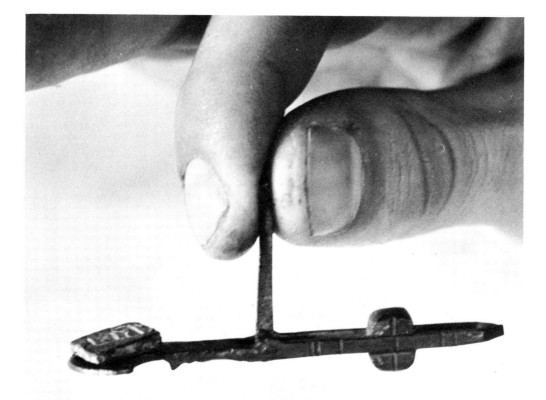

Fig. 26. Byzantine coin scale of steelyard type in equilibrium with its equivalent coin.

Fig. 27. Coin scale from Norway; eighteenth century. Note the sign on the beam at the left which is either a mastersign or an adjuster's mark.

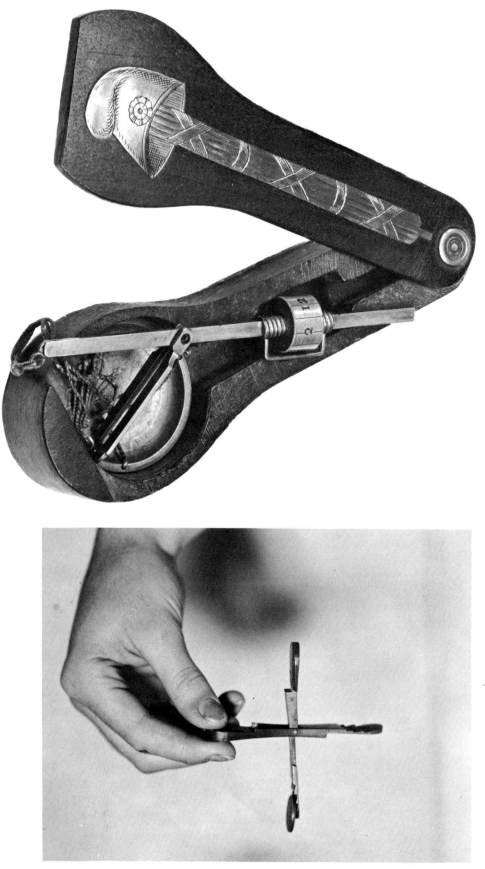

Fig. 29. French coin scale, constructed on the steelyard principle; end of eighteenth century.

Fig. 28. Modern coin scale using a principle very similar to that of Figure 26.

brass (there is a very nice seventeenth-century French example in the Streeter Collection). The big steelyards used today in Italy, Czechoslovakia, and other countries for fruit and in fish markets, butcher shops, etc. are usually made of iron.

Delicate scales of this type may even have beams of ivory, as is usual in the so-called Chinese opium scales (Fig. 25). They are of Chinese origin, but the name is not justified despite the fact that they may have been used incidentally for opium. The earliest accounts of them and their typical violin-shaped cases occur in the literature of the seventeenth century, where they are mentioned as fine scales for jewels and medicines. Leupold (1726) called attention to such a scale in the Museum of the English Society of Science in 1682 and reproduced it in Figure 3 of his Plate 19.

The counterpoise of the various steelyards was either a simple geometric form like a bowl, truncated cone, or cylinder, or it was an elaborate art form as was often the case in the Roman period. Figures 16 and 24 show counterpoises representing the helmeted head of Minerva and other personifications. The Streeter Collection has a late Roman counterpoise which is a bust of a woman. Occasionally the counterpoise was a fruit-like apple or pomegranate. In France in the seventeenth century sword handles (pommels) were frequently used by the scale makers as counterpoises. A large group of pommel counterpoises is in the Streeter Collection.

The counterpoise hanging on the long arm of the scale from an unattached metal ring could be moved along the arm until the beam was horizontal. The weight of the object that rested on the pan or hung from the hook on the other end of the beam (short arm) could be read immediately from a scale engraved on the beam. Leupold (1726, p. 57) mentions a scale invented by one Cassini which registered not only the weight but also the price of the goods, thus anticipating the scales in modern food shops.

The common method of moving the counterpoise was occasionally varied by incising a groove in the long arm of the scale, along which the counterpoise could be guided. A very nice example of this, a statera from the Byzantine period found in Egypt and now in the Museum of Turin, can be seen in Figure 26. The same type of money scale was used in Norway in the eighteenth century (Fig. 27) and is used in the Near East today (Fig. 29). Rarely, a small steelyard has a screw-nut instead of the usual type of counterpoise, which can be moved along the arm of the scale very slowly, thus permitting great exactness.

A late eighteenth-century French gold scale of this type is shown in Figure 29. The emblem (revolutionary cap and staff) on the original wooden container leaves no doubt about the period of its origin. This interesting and rare object is in the Wellcome Medical Museum in London.

65

In Roman times the simple steelyard or statera frequently had two modifications. One was a combination of the equal-arm balance with two pans and the principle of a steelyard. This was accomplished by using an additional rider* on one arm of the balance on which a scale was engraved. A rough equilibrium was reached by putting weights on one pan, and a fine balance was achieved by adjusting the position of the rider. This common device is found on a Roman beam without indicator and shears, probably from early imperial times, which was acquired by the author many years ago in Paris. Of the nineteen examples of balance beams that Lazzarini (p. 222) found in Roman museums, fourteen were of this type. A late revival of this ingenious combination of balance and steelyard is found in the riders on the beams of the analytical balances used today in every chemical laboratory.

The other modification of the steelyard in the oldest Roman specimens is a combination of the unequal-arm scale with the movable counterpoise and the principle of the changeable fulcrum as used in the bismar (see p. 56). This is very simply achieved by installing on the beam of a steelyard two or even three axes or fulcra. Because the same weight is used, the different length of the arm carrying the load requires a different weight-scale for each of the different fulcra, even if the counterpoise is in the same position and the beam exactly horizontal. Such scales are much more versatile, and most of the extant Roman staterae therefore have two or three scales engraved on the longer arm of the beam according to the different fulcra used by the merchant. This practical (probably Roman) invention was subsequently used fairly widely in Europe. The Chinese opium scales also often have two or three fulcra and, accordingly, two or three different divisions on the ivory beam, not engraved but marked by inlaid silver dots or other symbols.

A peculiar small scale of the steelyard type was shown to the author by Prof. J. Greenberg of Brooklyn, New York. It was intended for a physician who dispensed medicines. On the metal box in which it is kept is inscribed: Dr. C. H. Fitch's Prescription scale/Pat'd Sept. 29, 1885/Manftd by N. V. Randolph & Co./Richmond, Va.

SELF-INDICATING SCALES

The self-indicating scale has only one pan or hook for the load, and instead of a beam it has a heavy device suspended at an axis (fulcrum) and inscribed with a scale (Fig. 30). If the load is put onto the pan or suspended

---

* In Italian the rider is called *cavaliere* and this type of scale *bilancia a cavaliere*; in French the rider is called *cavalier*; in German *Reiter*.

from the hook, this substitute for the beam changes its position to balance the load automatically and an indicator points to the inscribed scale, immediately showing the weight of the load.

Leonardo da Vinci (1454–1519) is supposed to have invented this not-too-exact but very practical type of scale (Sanders, 1947). Later, mainly in the eighteenth century, the ducat scales (Fig. 12) combined features of the balance and indicator scales. The coin was put onto one pan and the exact coin weight on the other, or the other pan was made exactly one ducat-weight heavier. The shears immediately indicated that the coin was full weight or that it lacked a certain number of grains. This type of scale has repeatedly been revived and put to practical use, for example, by a minister named Philipp Matheus Hahn at Onstmettingen near Balingen (Germany), whose model of a self-indicating scale of the eighteenth century is preserved in the interesting Museum of Weights and Scales in Balingen, established by W. Kraut, founder and president of the great Bizerte scale factory there. A similar scale for hydrostatic weighing was constructed in Augsburg in 1760 by G. F. Brander (Antonowitz, 1957). Another handy self-indicating balance (pendulum type) was patented in London in 1863. Sanders shows it, and a specimen from the Streeter Collection bears the engraved inscription "PARNELL London" and "Halls Patent" (Fig. 30).

We quote here Sanders' short and clear description of this kind of instrument (p. 32):

A simple pendulum balance, patented in 1863, consisted of a pendulum made from flat sheet brass and suspended from a small shackle. The load was applied by a cord and ring. The circular pendulum weight formed a graduated scale at the centre of which was freely pivoted a counterbalanced pointer, which always hung vertically. When a load was suspended from the instrument the pendulum swung outwards to a position of balance, and the pointer, remaining vertical, indicated the correct weight. These little balances were made for the use of anglers, and also for weighing letters and postal packets.

SPRING SCALES AND TORSION SCALES*

The origin of the spring balance can be traced to the seventeenth century. Weigel (1698) describes it as "scales, newly invented made from steel working without pans and weights by a wire spring [*Drahtfeder*]."

Bion (1709) calls this kind of scale a *peson à ressort*.

---

* German: *Federwaage*.

67

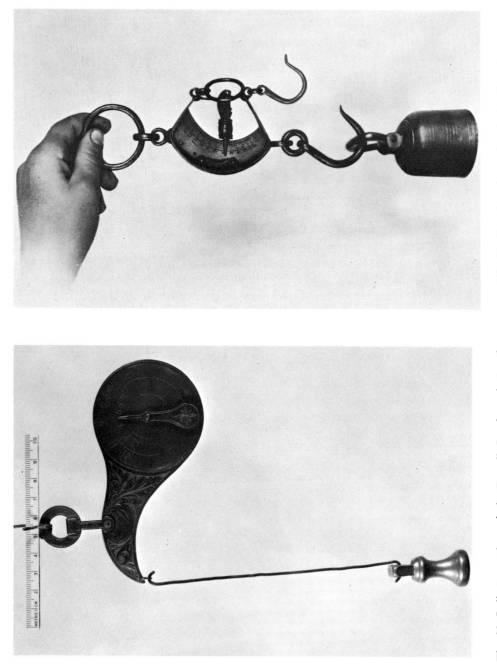

Fig. 31. French spring scale; nineteenth century.

Fig. 30. Indicator scale made by Parnell, London, nineteenth century, and patented. Pendant from the scale is a typical bell-shaped English weight.

The different kinds of spring scales, including the modern torsion scales, are based on the principle of deformation of any elastic material by the weight of an object attached to it, and measurement of the deforming power by the degree of deformation produced. The elastic material to be deformed may be a coil of steel or other material which is compressed by the weight of the attached object. The degree of deformation is indicated by a pointer on a division scale which is part of the instrument. Another type of spring scale is shown in Figure 31. The C-spring balance, much used in Europe as well as in the United States in the nineteenth century, is usually called the Mancur balance. The elliptical spring balance, in which the load deforms an elliptical steel ring, stems from the nineteenth century.

In the torsion scale an elastic band (for example, steel) or a rod is twisted on the place of impact of the weight. The amount of torsion corresponding to the amount of weight is shown by an indicator attached to the rotated material, revealing on a scale the amount of deformation.

Money Scales*

Weighing gold or silver coins has been a general practice from Roman times. Coin scales were very popular from about 1500 to 1800, but progressively less frequently used during the nineteenth century. In medieval ages (pp. 6 ff.) various laws in different countries prevented or required the use of coin scales.

In England in 1292, Edward I ordered in the statute "De moneta" that money should be weighed to prevent the circulation of pieces that were light or fraudulently clipped. For this purpose an apparatus called a tumbrel was to be used. Unfortunately, no specimen or picture of this interesting medieval English coin scale is known (Board of Trade *Report*, 1873).

Scales for this purpose had to be adequately sensitive and also as a rule small enough to be easily transported if a merchant wished to take them with him when he traveled. Since Byzantine times there have been especially designed containers of metal or wood for carrying scales and weights safely in the pocket (Fig. 32). From the Middle Ages to the nineteenth century manufacture of these containers in some centers of commerce (Cologne, Amsterdam, Paris, Lyons, Nuremberg, etc.) constituted a remarkable branch of the local industry.

The balances (equal-arm scales) among the money scales were as a rule remarkable only for their small size and the form of their pans. Only rarely, as in the case of the Nuremberg ducat scales, was their beam of special type

---

* French: *trébouchet, biquet*. German: *Goldwaage*.

69

(Fig. 12). In some places, such as Lyons or Paris, the pans of the small money scales were identical round basins. The mastersign of the balance maker, or the hallmark of the place where it was made, or both (Fig. 18) were usually impressed in the center. In Holland and Germany the pans of the money scales often were equal in weight but different in form. The one for the coin was a flat metal triangle, the one for the weights a round basin. Extant specimens of money scales having two triangular pans are of great rarity. One can see such a scale in Hans Holbein's picture of a young Danzig merchant (1532) and also in Marinus von Reymerswaele's portrait (1538) of the money changer and his wife, in the Prado Museum in Madrid.

The steelyard types of money scales were very common. Some of them have been mentioned already and can be seen in Figures 26–29. The scale in Figure 26, which I saw among the treasures of the famous Museum of Turin (Inv. No. 6353) is of special interest. It was excavated in Egypt and is well preserved; the counterpoise is of the sliding type, the steelyard easily movable in the fulcrum, and the weight is in the beam. That this was really used as a money scale for different coins can be proved by putting a Byzantine weight with the inscription $\beta\pi$ on the platform on the shorter side of the beam; equilibrium is reached when the center mark of the counterpoise stands just over one of the lines engraved on the longer arm of the beam by the balance maker (the β stands for 2, the π in this case apparently for scripulum). The money weight (Inv. No. 3666/3) proved to weigh 2.28 grams. The legal weight of two scripula is 2.27 grams. This corresponds to the weight of one half a *solidus* or *nomisma* (4.55 gm) as introduced by Constantine I in A.D. 307 and generally accepted as gold coin after the death of Licinius (324).

Unfortunately, the meaning of the other lines on the beam has not been ascertained, but Figure 26 proves that this object was used as a money scale as well as for weighing gold.

The weight of a gold *semis* was exactly 2.275 grams, and this actually existed as a gold coin from the fourth century (Ulrich-Bansa, 1949). Karl Pink (1938, p. 99, No. 93) even mentions an extant exagium weighing one half a nomisma. Figure 32, showing a Byzantine weight and scale box from the Flinders Petrie Collection, makes it clear that in Byzantine times equal-arm balances as well as steelyards were used as money scales, and that was true up to the nineteenth century, with a preference, of course, for equal-arm balances (see Figs. 88–91).

Mention should be made of one other type of money scale, very popular especially in England in the eighteenth and nineteenth centuries and in a different form in the Near East. This combines a steelyard and the principle of the ducat scale with pans of different weights. The long arm of this steelyard is heavier than the short one. If a certain coin of regular weight for which

70

Fig. 32. Byzantine scales and weights with container; from an Egyptian excavation.

Fig. 33. Collapsible guinea scale in mahogany container; eighteenth century.

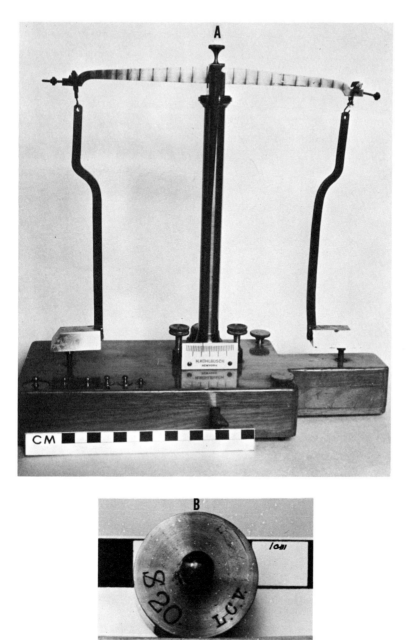

Fig. 34. (A) Large United States dollar scale with weights, made by H. Kohlbusch in New York; nineteenth century. (B) The twenty-dollar weight for this scale. The letters stand for "least current value" (Kohlbusch, p. 24).

the scale was made is put onto the platform hung from its short arm, the beam stays exactly horizontal, and the scale indicates equilibrium. If the coin is underweight, the beam will not remain horizontal, and the coin falls off. Figure 28 shows a coin scale like this found in Istanbul, which was adapted for use with various coins. This kind of scale was popular up to the turn of this century in Turkey and in other countries of the Near East.

The same principle but in different form (Fig. 33) was used as a collapsible coin scale in England. The beam usually had at its end a hinged weight, which was called the "turn." According to the direction of the turn the scale was adjusted to weigh guineas or half guineas. The entire apparatus was mounted in a mahogany box which could easily be put into the pocket. The collapsed scale was automatically set up when the case was opened. Sanders calls this the "spring-into-position" type.

A balance of American origin is depicted in Figure 34A.* It is of respectable size and was used exclusively by banks in this country for weighing gold dollars. Its most striking feature is the coffin-shaped pans, which in this case are not pans but platforms for the coin (right) and a coin weight (left). The denominations engraved on the weights are $1, $2.5, $3, $5, $10, and $20; their respective weights in grams are 1.4+, 4.0−, 4.6, 8.1+, 16.9, and 33.2. The scale, apparently dating from the nineteenth century, bears the manufacturer's name, H. Kohlbusch of New York, but the $20 weight has the initials L.C.V., probably indicating the maker of the weights (Fig. 34B).

PLATFORM SCALES†

The need for weighing heavy goods, in addition to small masses of gold, silver, copper, and so forth, was an early development and led to the construction of very large balances or steelyards. Figures 5, 6, and 15 show that bulk weighing was done in much the same way in Assyria in the ninth century B.C., in Greece in the sixth century B.C., and in Nuremberg in the fifteenth century A.D. In 1718 Jacob Leupold constructed for the city of Leipzig a hay scale on which wagons with loads up to 5,800 pounds could be weighed easily, with a sensitivity of about half a pound. This famous hay scale was described and depicted in his book *Theatrum staticum* in 1726. The elaborate scales for bulk weighing were undoubtedly the forerunners, if not the first representatives, of our modern platform scales. Of course they were of the steelyard type and did not employ the modern decimal and centesimal system.

* This specimen is preserved in the Carnegie Museum in Pittsburgh. I owe the photograph and all details to the kind cooperation of the Assistant Director, James L. Swauger.

† German: *Brückenwaagen*. French: *pont-bascule*.

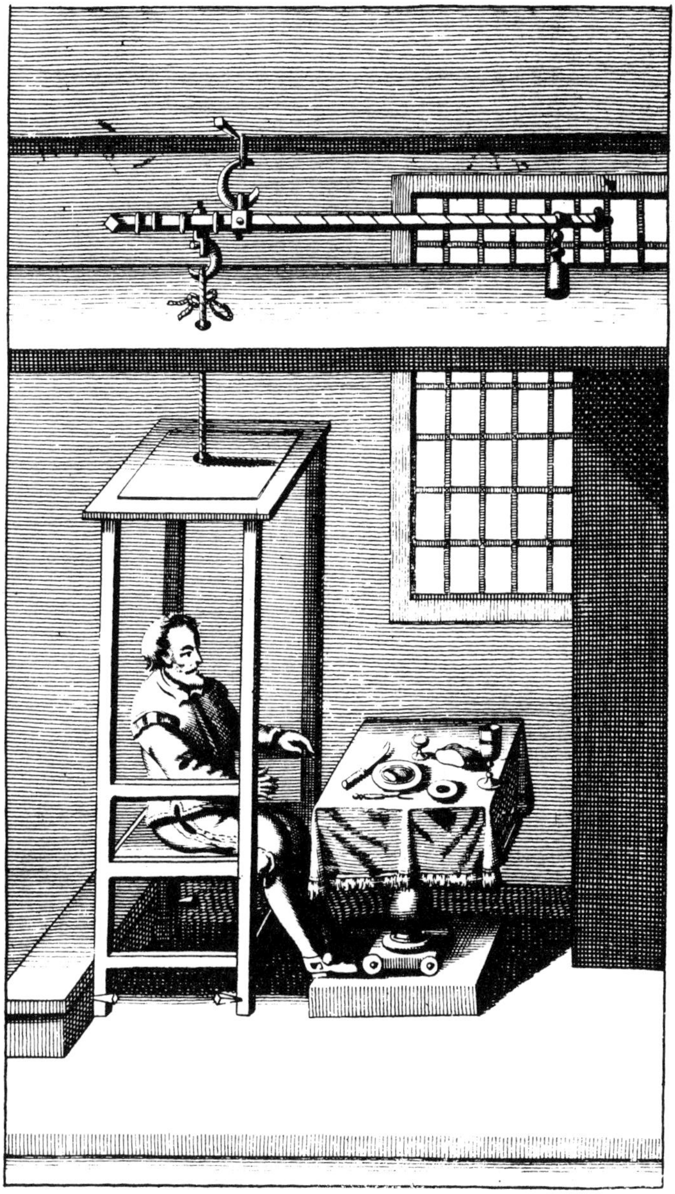

Fig. 35. A steelyard to weigh human beings for research purposes,
published by Santorio Santorio in 1614.

The best known of these special scales for large masses was the one built with a kind of giant steelyard by the famous physician Santorio Santorio (1561–1626) in Padua to weigh human beings (see Fig. 35). Leupold describes an improved similar scale as a *machina anthropometrica*. Platform scales are actually a balance based either on the principle of the steelyard or on a more or less complicated combination of levers for weighing heavy loads like carts and carriages; the balance pan carrying the load is a platform resting on the end of a lever. This type of scale is not so old as the spring scales invented in the seventeenth century (see p. 67). Sanders (1947, p. 19) traces their beginning from the English Turnpike Act of 1741 authorizing "all road trustees to erect at the Toll Gates any crane, machine or engine, which they shall judge proper for the weighing of carts, waggons or other carriages." According to Testut (1946, p. 36) and Sanders, John Wyatt, who died in 1766, apparently was the first to build such a weighing machine in England. Wyatt is said to have been employed by Matthew Boulton's firm, the Soho Manufactory (he also invented a yarn-spinning engine). A first weigh-bridge of his invention was erected at Birmingham Workhouse in England, but no details or exact data of the principle employed seem to exist about this invention or about the more complicated platform-weighing machines of Eayre and Yeoman in England (Sanders).

The first English patent for a compound lever and platform scale was given in 1833 to Thaddeus and Erastus Fairbanks, two years after they had been granted such a patent in America. However, the first inventor of a platform scale of whom we have exact knowledge was an Alsatian, Alois Quintenz, the son of a watchmaker in the little town of Gegenbach. Inspired by the new metric-decimal system he constructed a centesimal platform scale, for which important invention he received a royal patent in 1822. He died in the same year. His friend and financial sponsor Friedrich Rollé then persuaded the famous watchmaker Jean Baptist Schwilgue (born in Strassburg, 1776) to continue building and perfecting Quintenz' scales (Schmelzer, 1940, p. 27 ff.). This was apparently the beginning of the enormous development of this branch of scale making in Europe and all over the world. Schwilgue also became famous for his restoration of the ancient astronomical clock in Strassburg.

SCALES AS SYMBOLS

In ancient times the scale entered the vocabulary of mankind in a symbolic way and subsequently invaded painting, the graphic arts, and sculpture. The balances in the depiction of psychostasis in the Egyptian books of the dead indicate the use of this symbolism in Egyptian times. From the beginning of

76

the Christian era the same concept is constantly repeated in painting and sculpture. Justice, through the centuries and up to modern times, is always represented by a balance in the hand of a judge. It seems remarkable that for symbolic purposes an artist invariably uses a balance with two pans and an equal-arm beam. This is true of early examples—for instance Juno Moneta, represented on Roman coins with a scale in her hand—and it is also true in the medieval portraits of St. Michael and in modern art. A *statera* (steelyard) is almost never shown as a symbol, despite the fact that it was always a popular scale and known to everyone in the marketplace.

The reason for this probably is that the equal arms of the balance represent more clearly the idea of equal treatment. The two sides of the steelyard are of impressive inequality and, therefore, do not symbolize impartiality. The fact that the balance is more exact and sensitive than the steelyard may also play a part in this preference, but it seems less likely. Figure 36 shows an extremely rare example of a steelyard used for symbolic purposes. In this case the symbol does not represent justice but the burdens and responsibilities of a

Fig. 36. Medal bearing a rare symbolical representation of a steelyard.

prince, Frederic I of Gotha-Altenburg (ruled 1680–91). The only painting known to this writer in which the symbol of justice is represented by a steelyard rather than a balance is Sir Joshua Reynold's "Justice," which hangs in Oxford at New College, which had awarded Reynolds an honorary doctorate (Sanders, p. 6).

5. WEIGHTS

A weight is the materialization of the idea of a standard unit of mass. The criterion in commerce from earliest times has been that a weight of a certain type and standard be accepted as such by buyer and seller. The Bible, for instance, reports that Abraham bought the cave Machpelah as a burial place from the Hittite Ephron (Genesis 23: 16) for 400 silver shekels and that Abraham used a scale to weigh the silver, employing shekel weights that were recognized and used by the merchants. This indicates that certain kinds of weights were generally, in this case even internationally, accepted, and others apparently were not. Accepted weights must have been recognized by their form or material or by inscriptions or hallmarks.

The Hittite relief in Figure 3 shows a man with a small scale in his right hand and an object in his left hand which may have been the pouch in which the weights were kept. The Bible admonishes time and again that no one should possess two types of weights, to discourage the merchant from using a heavier weight for buying and a lighter one for selling.

Certain weights were used interchangeably in general commerce, as is done today, but other weights, that were used for particular goods only, were recognized by their form or inscription. In Egypt of the Old Kingdom certain stone weights of the *deben* standard were inscribed as gold debens (a weight unit). Pharmaceutical weights, the carat weight of jewelers, moneyweights, and the weights used with the Dutch corn scales are still known by their shape. Leupold (1726, p. 77) mentions the various types of weights used in Germany in his day for diamonds, gold and silver, mineral assays, coins, butter, meat, and pharmaceuticals. Whether these weights differed in standard as well as form Leupold does not make clear, but the latter is more probable and can be corroborated from other sources (see p. 170).

STANDARD AND COMMERCIAL WEIGHTS

Materials for Weights

Weights must remain stable. This is a postulate of commerce as well as of science; therefore material not subject to decay or chipping has always been sought. History has proved that stone and glass are the most satisfactory substances for weights, for metal is liable to oxidation and to corrosion which diminishes it. This fact, emphasized in the Talmud, was confirmed by the famous archaeologist Flinders Petrie (1926, p. 3) who said in connection with Egyptian metrology: "For the study of the units only stone weights should be employed." The materials offered by nature for weights are stones and the seeds of certain plants; both have been used in most countries of the world as far as we can judge from extant specimens.

Small plant seeds have an astonishing uniformity of weight, for example the carob seeds of the Near East, from which the carat weight used by jewelers derives its name. In Central Europe the dry grain of wheat was another natural weight, which gave name to the standard unit of one grain. This once widely used unit is now obsolete in all countries that accepted the metric-decimal system.

Babylonia, Assyria, Egypt, Judea, India, Peru, and even early Rome used stone weights. In Pompeii before A.D. 79 weights made of metal (brass, lead) were also in common use. The weights in classical Greece were made of lead, rarely of brass.

The kind of stone used depended on the natural sources at hand and also on experience with the nature of the stone—whether it was brittle or subject to decay. Its specific weight was also important. Apparently hard and heavy stones were preferred. In ancient Syria, according to Petrie (1934), hematite was the usual material for weights. In Babylon it was agate and hematite, onyx, "and other hard material" (Layard, 1853). For big weights (30 *mnas*) of the trussed-duck type, green basalt and white marble were also used (Brandis, 1866). The excavation of Layard in Nineveh prove that as early as the Babylonian period brass as well as stone was used for weights. Fifteen of the twenty-eight weights he found in Nineveh were in the form of recumbent brass lions. The others were of stone, all of the so-called trussed-duck type, a form typical of Babylonia and Assyria (Figs. 46, 74). All these were standard weights bearing inscriptions.

Weights in ancient Egypt were made from such stone as basalt, granite, limestone, syenite, alabaster, sandstone, diorite, hematite, serpentine, and porphyry. The weights in the form of animals or animal heads (Fig. 77) were of bronze, sometimes bronze filled with lead (Weigall, 1908). In ancient India, weights were spheres made of granite, pyroxene, diorite, or chert. For adjustment, holes were drilled into the spheres; in some of them the holes are still filled with lead (Marshall, 1957, p. 508).

Stone as a material for weights was later replaced everywhere by the heavier, less brittle and less vulnerable, but often less stable metals. The greatest advantage in metals appears to have been their high specific weight and consequently smaller volume. For example, in the author's collection are three Roman weights of the same form—the truncated sphere as shown in Figure 44. Two of them are three-ounce weights of serpentine, one with a diameter of 40 mm, the other 38 mm; the third is a six-ounce weight of brass whose diameter is only 32 mm. Thus it is understandable that the Greeks, for example, used lead for weights. But this very heavy metal oxidizes easily and is very soft; therefore the standard weights kept in the Greek temples for reference were usually made of bronze. The weights found in Olympia are an exception; there the material generally used was bronze (Pernice, 1894, p. 5). Marble was also used by the Greeks for large weights.

The Romans used stone for weights up to the third century A.D. (Pink, 1938), preferably dark serpentine. From the smallest fraction of an ounce up to fifty-pounders, they were nicely polished. The Streeter Collection includes a remarkable series of them.* The heavy Roman stone weights had simple or very elaborate handles of lead, remains of which can still be found in their sockets (see Fig. 45). The Streeter Collection possesses one interesting heavy weight with a handle of a light limestone, which is supposed to be of Byzantine origin (Fig. 70).

Besides the beautiful dark serpentine with its green spots and veins, other stone was used in the Roman realm and Roman provinces. Pernice mentions marble, limestone, and hard, green-black nephrite. Pink (1938) also lists basalt, travertine, serpentine, steatite, and tufa. According to this author, bronze was generally in use in Rome from the first tetrarchy. The treasures from the excavations at Pompeii† contain weights from A.D. 79 and earlier. They are made of serpentine or bronze, and some, owing to Greek influence, are of lead.

Bronze was the usual material for weights in the Byzantine period; only occasionally was glass or lead used. The practice of fraudulently reducing metallic weights by placing them in salt or a salt solution is mentioned in the Talmud (Baba Bathra, 90 a) and was strictly forbidden; also forbidden was the use of weights made of tin or lead or an alloy of other metals called "gistron"; only stone and glass were listed as permissible materials (Baba Bathra, 89 b). Pink mentions (p. 16) a Roman glass weight (before A.D. 300) in the Vienna collection, and there is also a well-preserved one in the Archaeological Museum in Jerusalem‡ that dates from the Byzantine period.

In the Islamic countries glass weights were commonly used in medieval times, mainly as money weights, and were very accurately made (see p. 8). The oldest example of such a weight was apparently issued by Qurrah ibn

---

* The fifty-pounder and a marked ten-pounder, both of serpentine, were acquired by the author in Naples and were probably from Pompeii. The others in the Roman series were acquired by Dr. Streeter in different places in Italy.

† I was able to check this, thanks to the kindness of Prof. Majuri, director of the Museo Nazionale in Naples.

‡ It is a greenish glass disk with a patina, 37 mm in diameter, 5 mm thick. On it in relief is the inscription NΓ; i.e. 3 nomisma. Its weight was 14.92 gm (4.97 gm per nomisma, whose official standard was $\frac{1}{72}$ of a Roman pound = 4.55 gm).

Sharik (709–714) (Fahmy, 1957); the Streeter Collection possesses a great number of them. Early Islamic coin weights of bronze were rare. One issued by the Umajad Viceroy Hajjag ibn Jussuf (A.D. 714) has been reported by Walker (1935, p. 246).

Even though late imperial Rome and Byzantium preferred brass for weights (in Byzantium usually inlaid with silver), exceptions certainly existed. The Talmudic prohibition against tin and certain alloys for weights proves that in the Roman and Byzantine periods these metals were actually used, because the Babylonian Talmud was codified in its present form in the sixth century A.D. I cannot agree with the opinion of Pink, whose thorough knowledge of the material is otherwise overwhelming, when he asserts (p. 16) that no Roman weights made of iron exist. They were extremely rare, but the Streeter Collection possesses one Roman eight-ounce weight, a genuine truncated iron sphere. The Byzantine weights were occasionally made of lead. One of these, a disc with a fine genuine patina bearing an inscription in Greek letters with a star on top and a cross in the center, is in the author's collection; its diameter is 41 mm. Kubitschek (1892) also mentions that he found among Byzantine weights from Dalmatia a quadrangular iron one-pound weight with typical inscription.

In a later period Egypt also turned from stone to brass for its weights. The extant examples from the Old Kingdom and up to the time of the New Kingdom are mostly of diorite, a dark-gray material, but other stone was used as well. The Streeter Collection possesses an interesting large scarab of a reddish porphyry, which is supposed to have been an Egyptian weight. It was apparently under Greco-Roman influence that Egypt changed later from stone to brass for weights, but she continued to use the typical cupcake form, so common in the New Kingdom.

Finally the Islamic countries also accepted the truncated spheres of Roman form for their weights. Occasionally, though rarely, lead was also used for post-Byzantine Islamic weights, as can be seen from an interesting specimen of this kind in the Streeter Collection.

Certain objects excavated in Chanhu-Daro, India (Mackay, 1943), and supposed to be weights from the third millennium B.C., are made of stone. Specimens are depicted by Mackay (Plate 91, Figs. 27–32) from the collection in the Museum of Fine Arts, Boston. The stones are listed as red quartzite, gray sandstone, chert, and agate. These weights (truncated spheres and cones, and cubes) need further study; they have almost the exact weight of the Oscan pound used in early Roman times.*

---

* The Oscan pound weighed 273 gm (Pink, p. 14); the two largest weights from Chanhu-Daro are 273.59 gm and 454.77 ($= 2 \times 272.38$) gm. This coincidence of data and form certainly suggests historical connections.

That pebbles were also used as weights in India can be proven from the Chanhu-Daro excavations where examples weighing 27.19 and 27.68 grams have been found next to well-polished rectangular weights weighing 27.53, 27.66, 27.14, and 27.34 grams. These represent practically one-tenth of an Oscan pound.

Weights have rarely been made from two different metals, although there are some Viking weights made of iron or lead sheathed in brass. This was also done in the eighth and ninth centuries in England, as is shown in a famous Anglo-Saxon weight in the Streeter Collection (Fig. 56), which is made of lead covered with an ornamented surface of gilt brass. Rare Viking weights of lead covered with brass (Brøgger, 1921: Figs. 19, 28, 35, 36) have been found near Bangor on the northeast coast of Ireland. They look much like the Streeter Anglo-Saxon weight. Later and up to the present time, standard weights and weights for analytical scales made of various materials (brass, pewter, gunmetal, etc.) were coated with gold or platinum to prevent corrosion of the less resistant material.

In former centuries pewter was used not only for market weights but also for standard weights in France and elsewhere in Europe. The *kilogramme provisoire* and the first standard kilogram in France were made from this material (Wolf, 1882), but pewter, being a very soft material, never became too popular for weights. A rare specimen of nested pewter weights from nineteenth-century Vienna is in the author's collection.

In 1529 Emperor Charles V sent a deputation to the mint in Paris to ask for a specimen of a two-mark weight, to be used as a standard in the mint of the Netherlands. For this occasion King François I of France had three pewter standard weights made, engraved on one side with the arms of the king, on the other with the arms of the emperor. He kept one for himself, the other two he gave to the envoys of Charles V—one for the emperor and the other for Margaret of Austria, at that time governor of the Netherlands. The original standard weight, according to which the three copies were made, remained under lock and key—in fact under three locks and three keys—in the "Cabinet de la Cour des Monnais" in Paris, and each of these keys was kept by a different dignitary of the mint. It was the standard weight of 1350 against which, according to an ordinance of 1494, each money changer, goldsmith, auditor of the pharmacists and spice dealers, and scale maker (*balancier*) had to check his own weights (Paucton, 1780).

Probably the earliest common use of iron for weights was in the Scandinavian countries. There, in Viking times, weights of iron with or without a shell of brass were in wide use, as is attested by specimens found in Viking graves and now in museums in Stockholm and Upsala. Ratls (Islamic weights) made of iron are mentioned in the Arabic literature as early as the fourteenth century (Levy, 1938). Copper, lead, iron, and other material for weights were used in Cologne from olden times (Kisch, 1960b) and in France as well (Machabey, 1949, p. 72). The heavy market weights in Cologne and France were still made of iron in the eighteenth century (Jaubert, 1773, p. 199); the small weights were made of lead.

In the Scandinavian countries brass weights were also common in the late medieval ages (Kisch, 1959). Even big weights of 100 pounds were made of brass, for instance, in Amsterdam in the eighteenth century (*Commissions-bericht*, p. 109). In fourteenth-century England, standard weights were made, by law, from brass. A statute of King Edward III in 1340 required the treasurer to have made certain standards of the bushel, the gallon, and brass weights (*darreissne*), to be sent into every county where such standards had not been sent before (Board of Trade *Report*, 1873). In 1392 King Richard II ordered again that "the clark of the market" should have all his weights and measures made of brass. And again in 1494 King Henry VII ordered that standard weights and measures of brass be delivered to forty-three cities and towns of the kingdom. But not until 1835 were weights of lead, pewter, or other soft materials prohibited in England by an act of Parliament.

Brass weights are also known from the Near and Far East (Persia, China, Japan), from Russia, and from other countries (such as the so-called Ashanti gold weights from Africa, used up to the nineteenth century).

The eighteenth and early nineteenth centuries found many scholars experimenting to find the most appropriate materials for standard and commercial weights. Various factors had to be considered. First, of course, the substance had to be as stable and infrangible as possible. The expansion coefficient, which determines for each material the amount of its expansion or contraction due to temperature changes, had also to be taken into account.

Gold and platinum, for instance for the imperial standard pound of England (Chisholm, 1877), and other materials like pewter or brass with a cover of gold have been used for standard weights. For larger weights bell metal has also been used, for instance in old Bohemia (Sedláček, 1923) or cannon metal, gold plated (in England; Chisholm). The famous arch-standard, the "kilogramme des Archives" in Paris, is made from a platinum–iridium alloy (nine parts platinum to one part iridium). The original kilogramme des archives was delivered by Fortin in Paris on June 22, 1798, along with the standard meter. It was made of platinum and recognized by the law of December 16, 1799, as the true and definitive kilogram (Block, pp. 13 ff.). Some European cities returned to the use of glass weights in the early nineteenth century.* Glass has, of course, the disadvantage of being fragile, but

---

* Von Alberti (p. 187) shows six Saxonian glass weights made in Friedrichsthal, Saxony, in 1815.

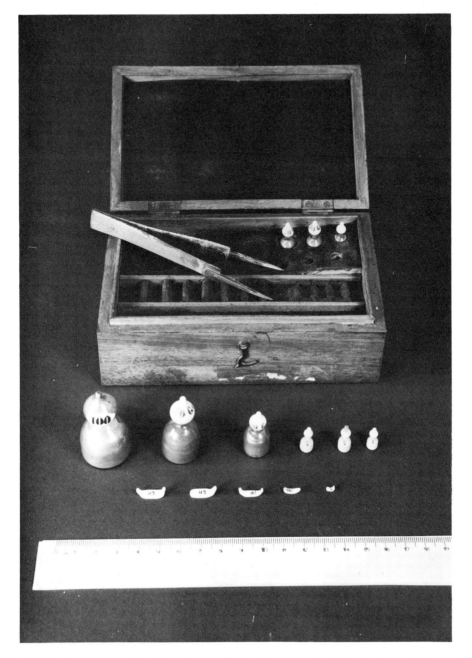

Fig. 37. Set of mercury weights (glass bottles filled with mercury) and simple glass weights; early nineteenth century Dutch.

Fig. 38. Glazed clay weight, Switzerland, 1669, weighing 1,185 grams.

Fig. 39. Standard iron weights bearing hallmark of Bern and a bear, representing the city's coat of arms; marked 1824.

Fig. 40. Standard brass weights of Bern from the eighteenth century, representing weights of 4, 2, and 1 mark (1 mark = 245 gm).

Fig. 41. Standard weights of Bern at right (1772) and of Basel. The one at the lower left weighs 250 grams (a quarter kilogram); the inscription "Muttergewicht" identifies it as a standard weight.

the great advantage of not oxidizing or corroding easily.

Figure 37* shows an interesting combination of glass and mercury used for weights. This method was tried in Holland in the early nineteenth century. The small units are glass plates bearing the denomination of their weight. The larger units are glass bottles which were filled with the appropriate amount of mercury, then completely sealed by melting the necks. Similar weights dating from the end of the eighteenth or the early part of the nineteenth century can be found in the Vienna Museum of Industry and in the Science Museum in London. In both these places the bottles are filled with small shot instead of mercury. In the Science Museum in London there are also larger cylindrical weights of the same period, made of solid, greenish glass.

Besides natural stones and glass, other fabricated materials have been used. Weights of glazed terra cotta were made in Switzerland up to the nineteenth century. Figure 38 shows a seventeenth-century pottery weight which is in the Historical Museum of Bern. Porcelain was also used; like glass, it is fairly resistant to oxidation and corrosion but is easily broken. Some years ago an antique shop in New York had an entire set of seven porcelain weights made in Birmingham by the firm of Avery. The Conservatoire National des Arts et Métiers in Paris owns a glazed white one-pound weight of terra cotta (No. 4845) with a black inscription: "1 lb à l'usage des charcutiers et des marchants du beurre" (one pound for the use of butchers and butter vendors).

Figure 39 portrays a beautiful set of nineteenth-century iron standard weights made in Bern; Figure 40 shows a set of brass weights made there in the eighteenth century. Both sets are adorned with bears, the symbol of the city of Bern. Similar iron weights bearing the stem of Basel can be seen in Figure 41.

The use of brass, iron, and pewter had spread all over Europe by the eighteenth century, especially for big commercial weights. The Streeter Collection possesses a unique set from France of twenty-three iron weights of 10, 25, and 50 pounds which stem from the seventeenth, eighteenth, and nineteenth centuries.

Standard metric weights in modern India are made from "admiralty bronze" (copper: tin: zinc, 88:10:2). They will be superseded later by stainless steel or nickel–chromium alloy (80:20). For smaller standard weights (500 to 10 mg), platinum is used; for 5- to 1-mg weights, aluminum is used (Aiyar, 1958, p. 16).

Thus scientists and governments, especially since the introduction of the

---

* I owe this picture to the kindness of the director of the Natural History Museum in Leiden, Dr. Maria Rooseboom, where the specimen is kept.

metric-decimal system, are interested not only in the metrological problems of exact standard weights but also in their physical properties.

A very instructive report of the different materials used for standard weights in France during the period of transition to the metric system is given in Wolf's paper (1882), which also lists the names and addresses of the first makers of the standard kilograms in Paris. Most of the European countries had taken good care of their centuries-old standard weights, for they were used only for checking from time to time, but the metric-decimal system made new standard weights necessary in the countries that accepted it. This created again the problem of choosing the best materials and forms. Some countries thought rock crystal was especially suitable; it has all the advantages of glass but is much less fragile and chipping does not easily occur. A standard weight of rock crystal is preserved in the collection of the Conservatoire des Arts et des Métiers in Paris. Consideration of the use of this material for a standard kilogram in the Austrian countries produced an interesting and extensive literature, published as the report of a special committee and called the *Commissionsbericht*, which was submitted to the Austrian Ministry of Commerce in 1870.

Before the rock crystal standard weight, Vienna—like other countries— used a platinum copy of the kilogramme des Archives, bought in 1857 and made by Froment in Paris (*Commissionsbericht*, p. 61).

A new type of standard kilogram of rock crystal was made by Dr. C. A. von Steinheil in Munich before 1837. He published two scientific papers on it (1837 and 1867). He carefully compared one of his crystal kilograms in 1837 with the kilogramme des Archives, which he called A^k; he called his own crystal specimen B^k. He sold B^k in 1846 as a standard to the Royal Government of Naples. Another specimen which Steinheil created was compared with B^k. He designated it Θ^k. On recommendation of the Viennese Academy of Sciences, Θ^k was acquired by the Austrian Ministry of Commerce as a standard kilogram (*Commissionsbericht*). Another standard weight of rock crystal can be found in the Science Museum in London.

The *Commissionsbericht* reports two other interesting facts concerning the new standard weights of this time. The standard kilogram of Berlin, made of platinum, had been brought there from Paris by Alexander von Humboldt. Dr. von Steinheil, who had created the rock crystal kilogram, also made for the mint in Vienna a standard pound of gilt brass (*Messing*). Another, electroplated in gold, was acquired in 1857 by the Viennese mint from the Prussian Normal-Aichungscommission in Berlin.

The importance of the density of the material of a standard weight is shown in the following example. The parliamentary standard pound of England is a cylindrical mass of platinum weighing 7,000 grains in vacuo (Clark, 1891,

p. 75). The commercial standard weight is of bronze; in vacuo it weighs the same as the platinum standard. If weighed in air, however, the platinum pound loses 0.403 grain, whereas the bronze pound, owing to its larger volume, loses 1.047 grains because of air buoyancy.* The standard bronze pound is therefore 0.644 grain lighter in air than the standard platinum pound (Clark). Facts like this had to be taken carefully into consideration by physicists and metrologists at a time when general laws concerning weights and measures were suggested for international acceptance.

The international conference on weights and measures in Paris (1872–75) (pp. 21 ff.) discussed also the question of material and form to be used for standard weights. In favor of rock crystal as the material of choice were its resistance against decay and corrosion, its ability to be perfectly polished, and its homogeneous density which ensured that all kilograms of this material would have the same volume at the same temperature. Despite all these merits, rock crystal was rejected mainly because of its low density. One kilogram of it would be about eight times larger than a kilogram of platinum and three times larger than a kilogram of pewter (Wrede, 1872). Further, the expansion of the volume of rock crystal in a range of 0° to 16° C is, according to Fizeau, about 50 per cent greater than that of platinum.

Gold has the advantage of not being liable to oxidation or corrosion, but it is too soft. It was thus decided (Wrede) that platinum—or still better, because harder—platinum–iridium should be the material of preference for standard weights. Gilt pewter was the second choice, and two standard kilograms of this material were acquired from Paris by Prussia.

The fact that bronze was also used for standard weights is proved by a set of plates of this material, which were Swedish (Stockholm) standard weights of 1804. They are now in the Museo Geographico in Madrid, in the valuable collection of standard weights of different countries assembled by King Carlos IV in 1804. In the same collection is a set of eight Russian standard weights (1804) made of iron.

Gurley's *Handbook of Weights and Measures for the Use of Sealers* (1912) discusses materials for weights in the United States. On page 24 it states that ordinary weights should be made of iron, painted or plated, of steel painted or plated, of brass or any substance of like hardness. Lead weights, unless encased in brass, copper, or like material, should not be allowed. As for small weights, Gurley informs the sealer that those of less than $\frac{1}{4}$ ounce or 10 grams should not be made of iron. Weights used by jewelers, apothecaries, and bankers (coin weights) should be made of brass, bronze, platinum, or aluminum. If

---

* The famous principle of Archimedes can be simply expressed as follows: "A body suffers an apparent loss of weight equal to the weight of the medium it displaces (water, air, etc.)." (Mettler *News*, p. 28.)

the sealer adjusts weights by a lead plug, it should be firmly embedded in the provided hole and be at least $\frac{1}{16}$ of an inch or 1 mm below the surface of the weight.

Shapes of Weights

Any product of human handicraft is given a specific shape by its producer. That has been true since man first created instruments for use in daily life— pottery, bows and arrows, axes and knives, and weights and scales. The shape of these tools is not dictated exclusively by their purpose. They bear witness also to the artistic level and manual skill of the craftsman. In addition, they reveal the personality of the artist and facts about his tribe and his generation, and in general illustrate the psychology of expression. In this regard, weights are very revealing, and for the scholar interested in the psychology of nations they offer a rewarding study.

Three basic shapes of weights exist: the geometric, the figurative, and the symbolic. Pebbles or stones in their natural state, without typical shape, must be omitted from this discussion, although they were probably among the oldest kinds of weights.

Geometric shape. Evidently the sphere was the first simple geometric shape to impress primitive man because he saw it in many fruits, and the circle was seen in the sun and at certain times in the moon. Baked and unbaked spheres of clay survive from the hands of the earliest potters. Spheres and such derivatives as hemispheres and truncated spheres are among the early geometric weights. Spherical weights are reported from ancient India (Marshall, 1957, pp. 209, 508). Hemispheres were used by the Jews (Fig. 42). Egg-shaped weights were found in ancient Egypt (Fig. 43), and truncated spheres (spheres with flattened poles) and truncated ellipsoids were used by the Romans (Figs. 44, 45). Other primitive weight forms are rectangular or cuboidal; there were also Etruscan weights of this shape, as a set in the Streeter Collection bears witness. Some of the stone weights found in Chanhu-Daro in India (see p. 82) are also geometrical—pyramids, cubes, and truncated spheres. Weights in Egypt during the Old Kingdom were chiefly rectangular blocks, usually made of diorite. Many of them bear typical engraved inscriptions. Truncated cones, most of diorite, with curved domelike tops similar in form to the American cupcake, were used during the New Kingdom in Egypt (Fig. 46).

Even in the eighteenth century (1767) the standard weight of one mark ($\frac{1}{2}$ pound) made in Vienna, was a truncated pyramid with a screwed-in knob. For heavy weights the rectangular brick form has remained a favorite up to

92

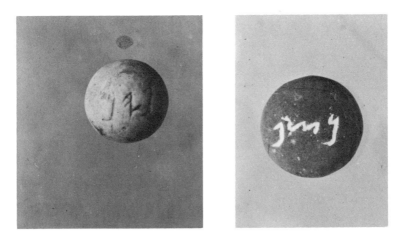

Fig. 42. Two Hebrew weights of reddish limestone, eighth to seventh centuries B.C. Left: (Inv. 4133) inscription "PIM" (8.2 gm). Right: (Inv. 3286) inscription "NESEF" (9.91 gm).

Fig. 43. Egg-shaped stone weight from Egypt with an inscription denoting 10 debens.

93

Fig. 44. Ten-pound Roman weight of the early imperial era; typical truncated sphere form of serpentine. The hole on top was originally fitted with a handle. This and the weight shown in Figure 45 were acquired by the author in Naples and probably originated in Pompeii.

Fig. 45. Truncated ellipsoid 50-pound weight of serpentine, probably from Pompeii, acquired by the author in Naples. The two holes on top were for the handle; the hole on the left still contains lead.

94

modern times. Figure 47 shows three iron weights of this shape that belong to the town scale from the old market (*Altermarkt*) in Cologne. One bears the inscription 1749 (Fig. 48), the year it was made. Handles on the sides of these iron hundred-pound weights were helpful in lifting them onto or off the scale. The similar centipondium of Vienna was made of brass (*Commissionsbericht*), and the same brick-like shape is typical of the heavy iron market weights from France in the Streeter Collection. They have a ring-shaped handle and date from the seventeenth and eighteenth centuries. The metric-decimal system was made obligatory on January, 1, 1840, in France. The form of the heavy weights was then changed, probably to prevent mistakes or cheating. The metric weights were hexagonal, usually a truncated hexagonal pyramid. Both the quadrangular and the hexagonal weights had an iron ring in an iron loop on the top to facilitate lifting.

Weights in the form of rings are also known to have been used in ancient Egypt, and ornamented ring weights in the form of flat round plates, made of brass and with a large central hole, were used in Persia and in the Near East from medieval times. The appropriate relative weights of a complete set make them easily recognizable as weights. The Streeter Collection possesses a

Fig. 46. Early oriental weights. Top row: three small Babylonian trussed-duck weights of hematite. Middle row: three Egyptian (New Kingdom) weights of diorite in cupcake form, photographed from the side and top. Bottom row (ends): two Egyptian weights of limestone, snake-head form (according to Weigall, 1908, probably imitating the trussed-duck form of Babylon). Between them is a barrel-shaped Syrian weight of hematite.

Fig. 47. A room in the Cologne Stadtmuseum im Zeughaus, showing in the foreground the old town scale, dating from 1615, and beneath it three iron brick-form 100-pound weights. On the wall behind it are beams from old scales, on top a wooden one from the eighteenth century. On the right wall are various brass measures of capacity, government standards from the sixteenth and seventeenth centuries.

beautiful set of four (Fig. 49). Their weights are 1167.5, 909, 604.5, and 302.5 grams (a ratio of 4:3:2:1). According to Dr. Streeter's note, these weights, which he purchased in London, had been found in Istanbul and were supposed to have been cast in China for Islamic traders. By coincidence another ring weight of 300 grams, supposedly from Persia and very similar to the four in the Streeter Collection, was found not long ago at a dealer's in Paris. It is now in the Smithsonian Institution in Washington.

The weighing unit for the five examples here described, which are now oxidized, partly corroded, and even perforated, seems to have been between 300 and 303 grams, or a fraction or multiple of this amount.

Hinz (1955, pp. 28 ff.) states that the weight of one *ratl* (an oriental weighing unit) in Egypt in Abassid's time (from A.D. 750) was 300 grams, in Fatimid's time (909–1171), 427 grams. The latter weight of a ratl of about 450 grams has governed the metrology of Egypt since the twelfth century. Hinz (1955, p. 31) also gives the weight of the *ratl rumi* in medieval ages in Asia Minor as 321.4 grams. Wherever in the Near East the five ring weights mentioned above may have originated, their weight of 300 to 303 grams corresponds approximately to the standard of Islamic weights of a very early medieval time.

In the Near East, ring-shaped weights were still in use up to the nineteenth century. One such specimen from the eighteenth century is in the Smithsonian Institution, and two nearly complete sets from the early nineteenth century (acquired in Israel) are in the author's collection.

In classical Greece the most popular form of weights was the quadrangular plate, adorned with symbols or figures of animals (turtles, dolphins, etc.) or man. This type and the disk form were commonly used later in Byzantium in continuation of the Greek custom. However, in Greece much more sophisticated weight forms were also used (Fig. 50). A weight very similar to that of Figure 50 is in the possession of the Museum of Jewish Antiquities in Jerusalem; it bears a Greek inscription. They also have a disk-shaped *mna* with a Greek inscription.

The Roman type of truncated-sphere weight found in Pompeii (Fig. 51) was especially popular in Islamic countries later, where the simple, truncated, double-cone form was elaborated by facets (very often pentagonal) bearing a birds-eye ornament (Fig. 52). Some of them can be dated from the impressed marks. The oldest known, as mentioned before, is from the eleventh century. Among the discoveries in countries of the Near East (Egypt, Israel, etc.) this form of small weight is fairly common.

Another specific Islamic weight is a cube with its eight corners cut by triangular planes. When this cuboid-octahedral form of weight (Fig. 53) was first used is not known. Examples have been found in Viking graves

Fig. 48. One of the centipondia seen in Figure 47, dated 1749. The wheel mark under the year may be a mastersign.

Fig. 49. Medieval Persian ring weights.

Fig. 50. Obverse and reverse of an ancient Greek lead weight, with various inscriptions.

Fig. 51. A set of bronze weights found in Pompeii, of truncated double-cone form.

(Kisch, 1959). The Vikings also used the typical Roman truncated-sphere form, which they adorned with pseudo-Islamic inscriptions (Fig. 54). In the Near East the cuboid-octahedral shape was in use up to the nineteenth century.

The disk and the polygonal forms were also used from medieval times up to the eighteenth century by different cities in southern France. Each city had its own easily recognized form of weight, usually distinguished by its coat of arms or some other symbol typical of the town (Gaillardie, 1898; Forien de Rochesnard and Lugan, 1957). Because of their shape (Fig. 55) these weights were subsequently also called *poids monetiforms* (Gaillardie, p. 4). Probably under Roman influence, disk weights were used in England in the eighth or ninth century, as is proven by the Anglo-Saxon weight in the Streeter Collection, and the shape was still being used during the reign of Queen Elizabeth I (Figs. 56, 57). However, the typical English bell-shaped weight was also used and has remained popular in England up to the present time (Fig. 58). It was also used for heavy weights in Venice in the eighteenth century (Fig. 59).

Fig. 52. Islamic bronze weight; a typical ornamented truncated sphere, the pentagonal facets adorned with bird's-eye decoration. The inscription on the top indicates that it is from the eleventh century.

Fig. 53. Oriental cuboid-octahedral brass weights used in Turkey up to the nineteenth century; acquired in Istanbul.

101

Disk weights were known in Germany in the fifteenth or sixteenth century; a beautiful five-mark weight (1,109 gm) is in the possession of the Cologne Stadtmuseum im Zeughaus (Fig. 60). Its provenance is not known, but probably it stems from the Rhineland or Westphalia.

Since the eighteenth century, cylindrical weights have been more prevalent in all European countries. These cylinders, as a rule, have a knob* on the top for easier handling.

The sophisticated geometrical weights used up to the twentieth century in Japan merit special mention. They are shaped like the body of a violin, as can be seen in Figure 67.

Fig. 54. Bronze weights from Viking graves, imitating the Roman segmented sphere. One is adorned with pseudo-Islamic ornamentation (twice natural size).

---

* French: *bouton*. German: *Knopf*.

Fig. 55. Obverse and reverse of a monetiform market weight of Toulouse (1239).

Fig. 56. Anglo-Saxon monetiform weight of lead; early medieval period.

103

Fig. 58. English bell-shaped weight of 1582, 56 pounds "averdepoiz."

Fig. 57. Eight-pound standard brass weight from the reign of
Queen Elizabeth I (1588).

Fig. 59. Bell-shaped artistically adorned weight from Venice, showing the year 1739 and the lion of San Marco; 50 pounds (24.384 kg; 27 cm high).

Among geometrical weights there are some of very odd shape, including the Islamic brass weights of truncated double-cone form dating from medieval times (Fig. 62). Figure 63 shows a typical small Venetian weight from the eighteenth century. Three of this rare type were bought by the author in Padua; they are of Venetian origin and were marked by a Venetian sealer. Figures 64 through 69 show typical specimens from the Castlereagh Collection in the Science Museum in London (see p. xv). They were used in 1818 in Lisbon, Barcelona, the Canary Isles, Cochinchina, Constantinople, and Tripoli respectively. Cushion-shaped brass weights from Algiers from the early nineteenth century are included in the collection of standard weights of King Carlos IV in Madrid (p. xvi). There are also many other types of weights which it is not possible to include here.

In the modern scientific approach to the desirable form for a weight the sources of possible weighing errors are considered first. Gurley's *Handbook* (p. 24) therefore warns the sealer that weights should have no sharp corners or angles that are susceptible of wear or breakage. It also warns against lodgments for dust or dirt. According to the same source (p. 24), weights larger than 5 pounds or 2 kilograms should be provided with handles screwed or driven in or cast without the use of lead or other binding material. In contrast to the modern ideal of weight forms, Figure 70 shows a complex Byzantine weight which is extremely liable to breakage and accumulation of dust.

An instructive summary of desirable forms for weights was given in the *Comptes rendus* of the Paris Academy of Sciences by Delamorinière and Séguir in 1857. Most countries today prefer the cylindrical weight form with a knob on the top as a handle. In laboratories, where exact weighing is imperative, touching the weights by hand is avoided by using a forceps. The small weights (less than 1 gm) used in modern laboratories are usually square plates of platinum or aluminum.

Standard weights in modern India are cylindrical, with smoothly curved knobs, rounded corners, and lightly engraved numerals. The larger standard weights have handles (Aiyar, 1958, p. 16).

It should be mentioned in conclusion that all weights can be damaged and invalidated not only by decay, corrosion, and other environmental influences but particularly by improper usage. This source of damage is, of course, especially important for standard weights. F. R. Hassler, the first director of the Bureau of Standards of the United States, therefore set up ten commandments for those who handle such weights. The ten points he emphasized still enumerate the greatest dangers today for a weight that is supposed to maintain its standard value indefinitely.

106

Fig. 60. Bronze weight of 5 marks (1,109 gm), showing St. Peter with a key, wheel ornaments, and a mastersign (arrow); probably sixteenth century Germany.

Fig. 61. Japanese violin-shaped bronze weight; eighteenth century.

107

Fig. 62. Medieval Islamic bronze weight of typical truncated, double-cone form.

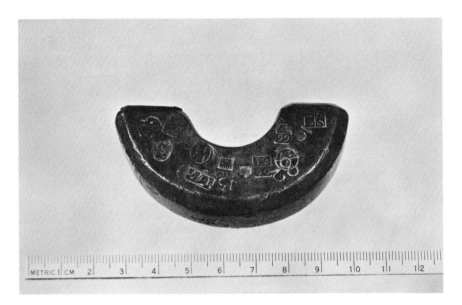

Fig. 63. Bronze weight from Venice, with sealer's marks; end of the eighteenth or early nineteenth century.

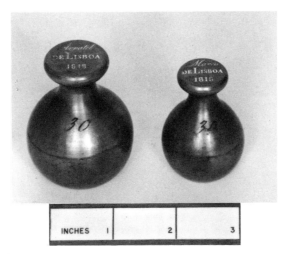

Fig. 64. Weights from Lisbon, 1818, the right one a mark the left one an arratel.

Fig. 65. Set of steel weights of typical form from Barcelona; early nineteenth century. Similar brass weights from the eighteenth century are in the author's collection.

109

Fig. 66. Iron weight of the Canary Isles (about 10 pounds); early nineteenth century.

Fig. 67. Iron weight of peculiar form, probably from Cochinchina; early nineteenth century.

Fig. 68. Set of domed octagonal brass weights from Constantinople; early nineteenth century.

Fig. 69. A set of weights from Tripolis (Greece); early nineteenth century.

111

Fig. 70. Byzantine stone weight of grayish limestone.

Figurative shape. The various weights described in the preceding section occasionally bear ornaments (see the Islamic weights, Figs. 52, 53). Figure 71 shows the geometric and other kinds of ornamentation that embellish the quandrangular weights of the Ashantis in Africa. Abstract ornamentation appeals alike to primitive artists and to those of a highly developed culture— witness the stylistic ornamentation in Egypt and the oriental arabesques. However, mankind has also tried to present the objective world in painted and sculptured images. These were used on religious articles as well as on objects of daily use such as instruments for weighing. Weight and scale makers used the images of animals and men; parts of plants such as the pomegranate and even parts of the animal skeleton like the famous knucklebone (*astragalus*) were used by the Greeks and by the Romans as weight forms. Figure 72 shows the typical astragalus form in a Roman weight from Pompeii of remarkable size. A very small, beautiful astragalus of brass from Greek times acquired in Athens (author's collection) has a weight of only 30 grams. Figure 73 shows a similar one also from Greece in the Kunsthistorische Museum of Vienna; its weight is 920 grams.

From the animals chosen to be represented by weights one may infer the varieties that were well known to the population of a certain area. As already mentioned, the most typical in Babylonian times are the so-called trussed duck weights which probably (but not necessarily) represent ducks; they could also represent geese (Figs. 74, 75). Other typical weights in the form

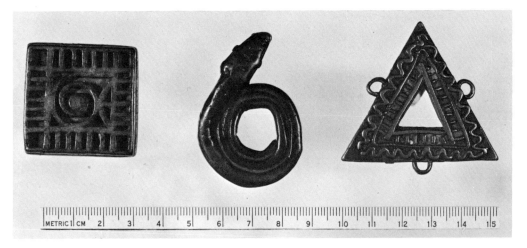

Fig. 71. Ashanti gold weights of brass (African Gold Coast).

113

Fig. 72. Large Roman astragalus (knucklebone) weight.

Fig. 73. Small Greek astragalus weight (920 gm).

Fig. 74. Large Babylonian trussed-duck weight from the reign of King Eriba (802–763 B.C.).

of lions (Fig. 76) were used, as indicated by the double inscription, in both ancient Babylon and Assyria. Egypt had weights in the form of rings and animals or heads of animals, as can be seen in pictures from ancient Egypt depicting scales used in commerce. One such weight in the form of a cow from the famous collection of Flinders Petrie is seen in Figure 77 (now in the possession of University College in London).

According to Petrie, Egyptian animal weights stem from the Eighteenth and Nineteenth Dynasties (1575–1194 B.C.) In the New Kingdom in Egypt (sixteenth to eleventh centuries B.C.) another weight form was a snake head (Weigall, 1908, Fig. 48); it is probably a modification of the Babylonian trussed-duck form.

This is perhaps the proper place to introduce Petrie's list (1926, p. 6) of the distribution of the different shapes of weights in the different periods of Egyptian history, according to the results of his excavations and studies. The years have been taken from Gardiner's book (1961):

Cylinders and cones with domed base	Amratian, prehistoric
Pointed cones	Semainian, prehistoric
	Dynasty
Round-top cones	18th (1575–1308 B.C.)
Square, sharp edges	1st
Square, "greatly rounded"	4th (2620–2480 B.C.)
Square, "edges less rounded"	9th
Square, "edges slightly rounded"	12th (ca. 1900 B.C.)
Oblong cylindric top	3rd, 12th, early 18th
Pillow	4th–12th (2620–ca. 1900 B.C.)
Black quartzose cube	22nd–30th (945–343 B.C.)
Domed top [Roman]	4th–26th (2620 B.C.–7th cent.)
Domed	26th–30th (664–343 B.C.)
Barrel [Syrian]	18th–23rd (1575–730 B.C.)
Duck [Babylonian]	18th–23rd (1575–730 B.C.)
Animals [Egypt]	18th–19th (1575–1194 B.C.)

It may be of advantage to add to this catalogue an abbreviated survey of the Egyptian dynasties according to Gardiner (pp. 433–47).

	Dynasty	*From about* (B.C.)
Old Kingdom	3rd	2700
	4th	2620
	5th	2480
	6th	2340
	7th	?

I. Intermediate Period

Middle Kingdom	11th	1991
	12th	

II. Intermediate Period; the Hyksos

New Kingdom	18th	1575–1300
	19th	1308–1194
	20th	1184–1087

Late Dynastic Period

	21st	1087–945
	22nd	945–730
	23rd	817?–730
	24th	720–715
	25th	751–656
Saite	26th	664–525
Persia	27th	525–404
	28th	404–399
	29th	399–380
	30th	380–343
	31st	343–332
	Alexander	332

Fig. 75. Two small duck weights found in Nimrud; from the eighth century B.C.

117

Fig. 76. Set of four Babylonian lion weights (in pounds:ounces/drams weighing respectively: 32:14/14; 11:1/14; 4:6/4; 8/5).

Fig. 77. Egyptian cow weight.

Fig. 78. Set of four goat weights from Pompeii.

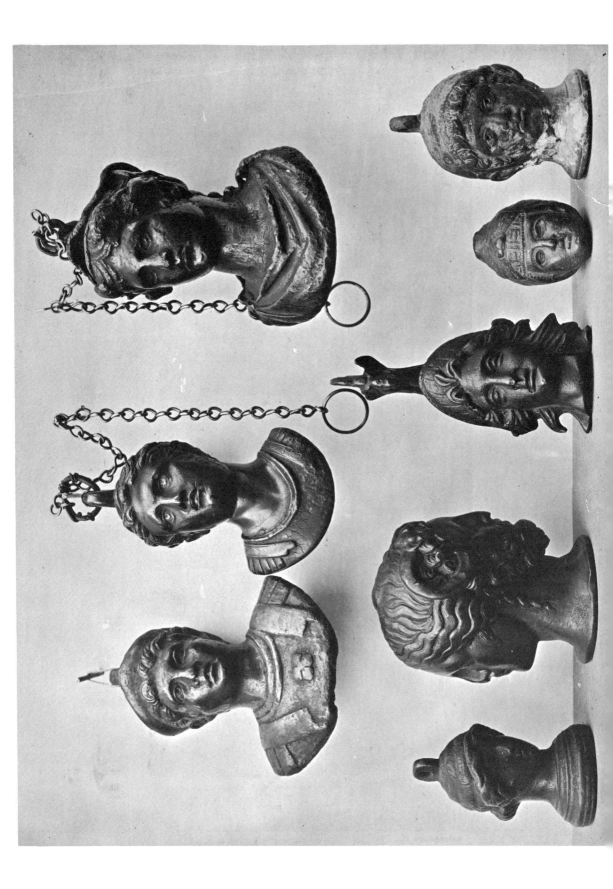

According to a Greek inscription (Pernice, 1894, p. 7) commemorating the agoronomos P. Memmias Agathocles, weights in ancient Greece not only took the shape of a knucklebone but also the form of animals, deities, and mythical characters (Eros or Atalanta). Unfortunately, only specimens of the knucklebone have survived.

It is different with the weights of the classical Roman period; the excavations of Herculaneum and Pompeii have brought to light a great number of well-preserved weights and counterpoises of steelyards in both geometric and the most diversified figurative forms. Of course, Pompeii was strongly influenced by Greek art. A large weight was found there in the form of a boar (a 100-pound weight) even bearing a Greek inscription (Pernice).

The Museo Nazionale in Naples has an entire set of six weights from Pompeii depicting goats (Fig. 78), which were probably the most common animals in the streets of Naples then, as they still were even in the early twentieth century; the goat gave the island of Capri its name. However, in the Roman Empire and its colonies (Pink) weights in the form of truncated spheres and double cones were most prevalent, as can be gathered from the many extant specimens (Fig. 51). The astragali and the figurative weights were a minority.

The counterpoises of steelyards, however, were generally decorative, and were sometimes masterpieces of art. Most depicted heads or busts of deities such as Juno or Artemis, or of unknown persons (Fig. 79). Sometimes a counterpoise was made in the shape of a fruit or a vase.

In other countries, too, some weights represented the more common animals; for instance, snakes in India, elephants in Siam, birds (chickens) in Burma, and so forth (Fig. 80). Weights in the form of horses (in bronze) were well known in the Scandinavian countries in medieval times (Figs. 81, 82) and have been the subject of modern studies (Brøgger, 1921; Falck-Muns 1939; Wideen 1953, 1954; Kisch 1959).

The Historical Museum of Lucerne possesses two remarkable old figurative weights. One, dated 1540, is a bell-shaped, 100-pound weight from Zurzach that is called the "Zurzacher Glockencentner" (Inv. No. 1039). The other, supposedly very old also (Inv. No. 720), cannot be dated exactly; it is called the "Ruschwihler Affengewicht" and represents a monkey with a falconer's satchel playing a kind of guitar.

Whether the charming little horses of about 1000 B.C. from Luristan (Near East) were used at one time as weights cannot be proven. Unquestionably the most numerous and diversified figurative weights are those produced up to the nineteenth century by the Ashanti, an African people living near the Gold Coast, the so-called Ashanti gold weights. They range from figurines of humans illustrating popular proverbs, to fish, birds, crocodiles, cockroaches, dragonflies, guitars, guns, and drums—almost everything that can be seen in the daily life of this tribe (Fig. 83). The Ashanti were not too particular about how they adjusted these figurines to the correct weight: they simply cut off half a leg or even an entire one. This people also had a very diversified group of geometric weights (cubes, pyramids, etc.; see Fig. 69).

Of the large number of figurative weights from different countries and nations of antiquity, only a few can be listed here as especially typical:

Babylonia	trussed duck, lion
Assyria	lion
Egypt	lion, cattle, snake head, gazelle, hippopotamus
Greece	astragalus (knucklebone)
Rome	astragalus; heads and busts as counterpoises

Jews and Arabs avoided making weights in the form of figurines, in accordance with scriptural interdicts. Only a single figurative Jewish weight, in the form of a turtle with an inscription, has been found. Its inscription gives its value as half a quarter (i.e. shekel) and, according to Reifenberg (1936, p. 34) it dates from the sixth to fifth century B.C. (see also A. S-N., 1963, pp. 865–66).

Symbolic shape. Like abstract art and objective reproduction of the surrounding world, symbolism has been a form of artistic expression from earliest times. Mention has already been made of the custom of the Ashantis to symbolize in their figurative gold weights proverbs known to their tribesmen. Weights of a symbolic shape also played a particular role in Dutch pharmacies in the seventeenth and eighteenth centuries and elsewhere in Europe, for to mistake weights in a pharmacy could have fatal consequences for the patient. We therefore find as early as the seventeenth century pharmaceutical weights whose shape unmistakably indicated the weight, and could guide the tyro in the pharmacy even if he was illiterate. More on this topic will be found later; only the pharmaceutical weights, as depicted in Figure 84, are of interest here. Their shapes are those of the weights as written by the physician on the prescription. For instance ℥ stands for an ounce and ʒ for a dram. The symbol of each denomination was identical with the shape of the respective weight and even the problem of expressing two scruples was easily solved, as shown in Figure 84.

NESTED WEIGHTS

For more than 2,000 years nested weights have been important in commerce, and they have a remarkable history of their own that warrants a separate discussion.

Fig. 80. Weights in the form of a chicken (Burma) and an elephant (Siam).

Fig. 81. Medieval Scandinavian brass weights in the form of animals, weighing 208.38, 197.28, and 206.75 grams.

Fig. 82. Medieval bronze weight from Scandinavia.

Fig. 83. Figurative Ashanti gold weights of brass (African Gold Coast).

124

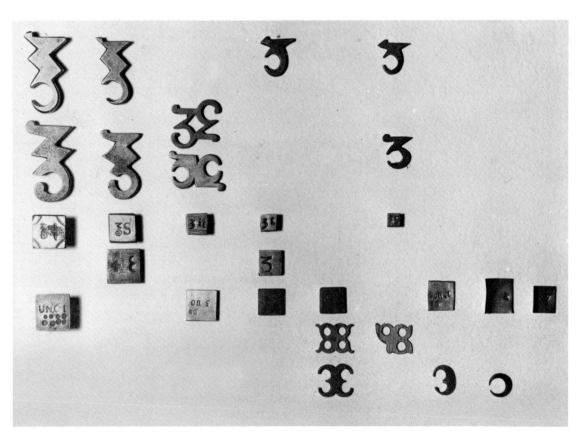

Fig. 84. Pharmaceutical weights from the eighteenth century.

Nested weights* are a series of weights, usually in the shape of cups, which fit into each other and therefore require a minimum of space. Figure 85 shows a set from a typical eighteenth-century Parisian gold-scale box. Very early examples of such weights are rare. Those in the Streeter Collection that date from Roman or Byzantine times are somewhat crude in workmanship and the cups do not nest well. Most of them stem from the late Roman or Byzantine period or, if found in Egypt, from early Islamic times. It was in the period from the sixteenth to the nineteenth century that their manufacture was brought to a high degree of perfection, and they played an important role in every kind of business as well as in kitchens and pharmacies. The largest cup usually had a hinged lid so that the set became an extremely handy, compact unit.

Nested weights were made in vastly different sizes. Some were so big that the weight of the entire set was 64 pounds or even more. Some weighed only 16 or 32 ducats (1 Cologne pound = 134 ducats). The art of making these weights demanded the highest craftsmanship. Their basic structure is as follows. The largest cup, which we shall call the master cup, weighs exactly the sum of all the smaller ones. The second largest cup weighs exactly half the master cup and the sum of all the remaining cups. This ratio continues from cup to cup, down to the smallest insert which is not a cup but a disk that weighs the same as the smallest cup and fits exactly into it. A set of nested weights is not complete without this small disk; it is the one most commonly lost, and often it has been replaced by a dealer or even a collector by a replica of later date.

From the sixteenth century (or even earlier) up to the eighteenth, the city of Nuremberg, famous for the high standards of its coppersmiths and craftsmen, had a monopoly on the manufacture of nested weights, which were exported throughout the world (Stengel, 1918; Borssum-Buisman, 1952; Kisch 1960b). It was only in the late eighteenth century that other countries learned the art of making nested weights (Kisch, 1960b).†

---

* German: *Einsatzgewichte*. French: *marc, pile, poids de godet*. Dutch: *sluitgewichten*.

† I have found recently additional confirmation of evidence (1960) that Nuremberg originally had a real monopoly on making nested weights. A scholarly book on money, weights, and measures in 1780 (Paucton) also describes (p. 53) nested weights. At this time in France they were simply called *marc*: "Ces sortes de poids de marc par diminution, se tirent tout fabriqués de Nuremberg; mais les balanciers de Paris & des autres Villes de France qui les font venir pour les vendre, les rectifient & adjustent, en les faisant vérifier & étabonner sur le marc original & ses diminutions pareilles gardés dan les Hôtel des Monnoies." (This kind of weights [de marc] in descending size are all made in Nuremberg. But the scale makers of Paris and other towns in France who import them for sale adjust them and have them verified and checked with the original marc and its smaller fractions which are preserved in the mint.)

We learn also from Paucton that the old original "marc" was made by order of "Roi Jean qui regnoit en 1350" and that it was of copper (perhaps brass is meant) "avec sa boîte de même métal" (p. 645). Indeed nearly all extant nested weights are of brass; rarely are they copper or silver. Only two nested weights of pewter are known to this writer. One was made in Vienna in the nineteenth century and is in his own collection, the other is in the collection of the Museum of the city of Prague. The master who made a nested weight, the country where it was used, and very often also the date of its origin can be identified by marks impressed on the cover of the master cup and sometimes also on the bottom of the smaller ones. There will be more about this in the following chapter on inscriptions.

Nested weights displaying their inserts or only the closed master box can be seen in early sixteenth-century paintings of money changers or jewelers. At this time they were very plain. The top of the master box was round and flat; the latch was simple and typical of this period. Two such sets are in the author's possession. One, acquired in Florence, has the mastersign of a Nuremberg coppersmith on the top, and inside each cup is the hallmark of Florence (the arms of the Medici) with the clearly legible year 1578, which proves the importation of nested weights from Nuremberg to Florence in the sixteenth century.

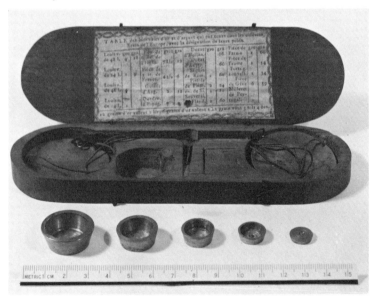

Fig. 85. French scale and weight box. In front are five cups from a nested weight that fit into the body of the box; eighteenth century.

Fig. 86. Typical Nuremberg nested weights. The closed master cups contain sets of weights; seventeenth century.

During the seventeenth century, the Nuremberg nested weights became more and more elaborate. The German Renaissance and baroque art styles are well exemplified in these little masterpieces. The master cup was now embellished with bands and by a carrying handle. Hunting scenes were engraved on the outside of it; the two columns to which the handle was attached arose from the cover of the master cup. These columns were either simple or adorned with mermaids, knights, or heads of knights. For the handle itself, for reinforcement of the cover, or as clasps to close the master cup seahorses or horses' heads were very popular (Fig. 86). One finds rarely an entire horse as a handle. Such handles were attached not only to the heavy sets for practical purposes but were also added to very small sets, apparently only as decoration. The most beautiful of the Nuremberg nested weights appeared in the seventeenth century.

In the eighteenth and nineteenth centuries, nested weights again became simpler. The handles, the mermaids, the seahorses, knights, and engraved scenes disappeared. Their increasingly common use in house and business necessitated a reduction in their cost and, therefore, a simpler design. Finally, the competition from manufacturers in other countries (Austria, England, France, Italy, Sweden, etc.) became more noticeable. England and America, with their more practical approach, usually omitted the master cup and its elaborate lid entirely in their designs. Simple sets of cups fitting into each other are still used in these countries.

In conclusion, a variant of the nested weights, the telescope weights, should be mentioned. The disks have an elevated rim, similar to a weight preserved in London from Elizabethan times (Fig. 57) and fit exactly into the rim of the next larger one. The smallest of them, as in nested weights, is usually a flat disk without a rim. Thus, if put together, the entire set forms a pile that looks like a collapsed telescope, which is probably responsible for the name. Such sets of weights, still used in England and India, usually do not include a master box. There is an exception; in Russia during the nineteenth century, where telescope weights were common, they were provided with a metal container shaped like a cupola dome, a typical shape which I have seen only in these Russian weights. The Museo Geographico in Madrid has a set of Russian telescope standard weights (1864) made of iron but without a container.

COIN WEIGHTS* AND WEIGHT AND SCALE BOXES

Since coined money came into use, probably in the seventh century B.C., it has been closely related to the weight of the precious metals [gold, silver,

---

* Latin: *exagia*. French: *poids monetaires*. German: *Münzgewichte*.

Fig. 87. (A) Six weights from late Roman, Byzantine, and medieval periods. Top right is a 2-ounce weight (the horizontal line before the Roman II is a symbol of ounce); the inscription reads: 2 ounces [equal to] sol[idi] XII. The two weights at the lower right are exagia solidi of late Roman imperial times, bearing the heads of emperors.

Fig. 87. (B) Reverse of the weights shown in (A). Top left, from the reign of Theoderic, inscribed as a 3-ounce weight. The figure with the scales on the exagia solidi is Juno Moneta.

131

electrum (elektron in Greek)] from which it was made. The monetary reform of Emperor Constantine the Great (A.D. 307) put the weight of the gold solidus at $\frac{1}{72}$ of a Roman pound; i.e. 4.54 grams.

Gold and silver were extremely valuable in early times, and two kinds of people have always been interested in the exact weight of circulating coins made from these metals: money changers and merchants needed to check the weights of the coins they accepted; sharp dealers looked for coins that were heavier than the legally permissible lowest weight (in German *Passiergewicht*), because of incorrect weighing at the mint, and then trimmed them down to this minimum and sold the surplus metal. That the practice of weighing coins was already popular in Byzantine times is proved by the coin scales from this period found in Egypt (Figs. 26, 32).

Delicate and sensitive scales and reliable weights were needed for exact coin weighing. Special weight units, known from Roman times as *exagia solidi* (from *exigere*, to adjust), were made for this purpose, each representing a particular coin. The exagia must have been used in Rome commonly in imperial times, since Emperor Julian in A.D. 363 permitted only official weight makers to possess and use them (Werner, 1954). The ducat scales, already discussed, checked the exact weight of coins in a balance having one pan exactly one ducat heavier than the other.

Standard weights of bronze for adjusting weights used in commerce existed in Greece where (as in Rome) they were kept in temples (Pink, 1938); later, under Emperor Justinianus (527–565) for example, they were kept in the main church in Constantinople (Barbieux, 1926, p. 18). Standard weights were also kept in the Roman mints and in the Roman military camps. In the Greek era certain government officers (the *zygostates*) were entrusted with the control of the correctness of weights and measures.

Especially noteworthy are the exagia solidi (Fig. 87), the coin weights to check Constantine's new gold coins, from A.D. 307 on. They were the first specimens of the great series that flourished up to the nineteenth century. Made first by the government, the privilege was later turned over to private balance and weight makers, whose products were checked and adjusted by an officer of the government and officially stamped with a government hallmark.

Each coin weight had to fulfill certain specifications in addition to being the exact weight of a certain coin. In times of widespread illiteracy and equally widespread use of coin weights by merchants, it was necessary that the weight for a particular coin be easily recognized. For this purpose the face of the coin was usually reproduced entirely or in part on the weight. Later, especially in Holland, labels depicting the different weights and the names of their coin counterparts were pasted inside the lids of the weight and scale boxes used by

132

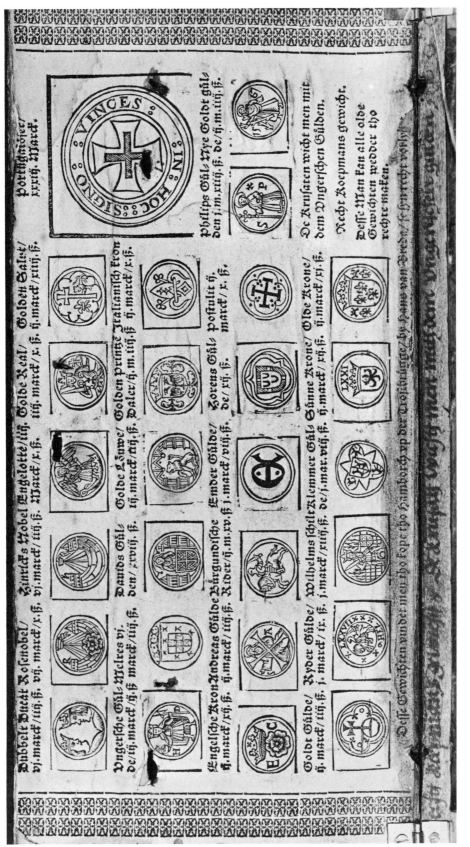

Fig. 88. Label in the lid of a weight and scale box from the sixteenth century (Hamburg, 1587) depicts the various coin weights in the box and lists the coin-weight equivalents.

money changers and merchants (Figs. 88, 89). Another important require-
ment of coin weights was that they be easily distinguished from the actual
coins. In ancient Rome both square and round exagia were used (Fig. 87).
In medieval France triangular and hexagonal coin weights were popular, but
later they were usually round or square. In England, first the hexagonal, then
later the round and quadrangular coin weights were commonly used. Square
coin weights were usual not only in France but also in Belgium, Holland,
Cologne, the Rhineland, Nuremberg, Venice, Turin, and Florence (there
with rounded corners). The round monetiform coin weights prevailed in
Prussia, Austria of the eighteenth century (Fig 17b), Milan, and Spain.

One way of preventing the misuse of round coin weights as coins was to
make them a different size or leave the reverse plain, to be used for the adjuster's mark. It was usual in Antwerp, Holland, and occasionally also in
Cologne, to put the mastersign of the weight maker and the issuing date on
the reverse of the coin weight. As a rule brass was used for coin weights; lead,
white metal, or silver was used only rarely.

From Byzantine times (Fig. 32) it was customary to keep the small, sensitive, and easily damaged gold scales (coin scales) and the various coin weights
together in a box. These weight and scale boxes (Fig. 89), usually made of

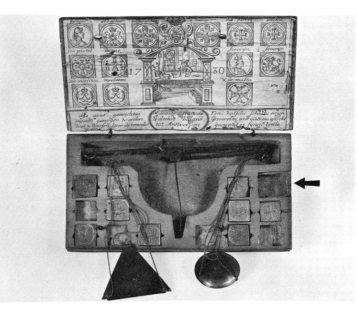

Fig. 89. Weight and scale box made by J. F. Wolschot, Antwerp, dated 1730. The label shows a man
using the gold scale and indicates which weight is for which coin. The gold weights are in their
recesses; the scale has one round and one triangular pan. The arrow indicates a small container
for grain weights (lid missing).

134

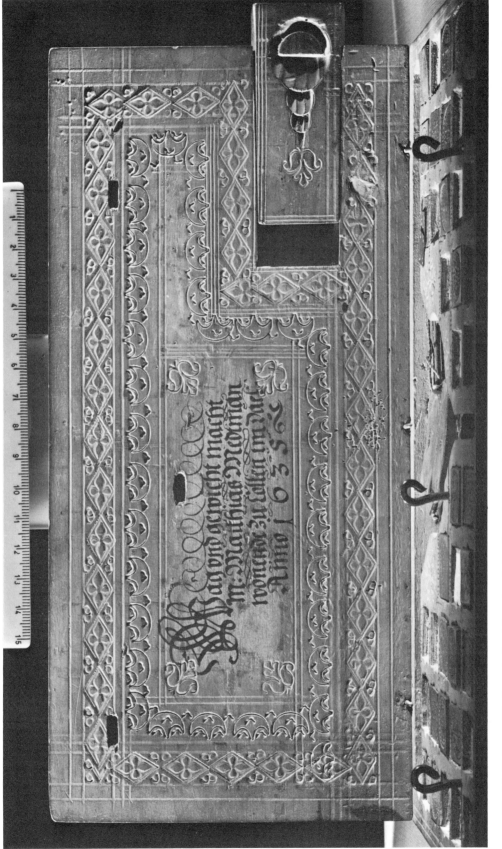

Fig. 90. Elaborate lid of a typical weight and scale box from Cologne. The lid of the recess for the grain weights is half open; three hooks in the body of the box fit into the three holes in the lid for locking the box.

Fig. 91. A typical Cologne weight and scale box from the early seventeenth century with scale, weights, and the inscription of the balance maker (Rütger von Essen) and his address.

wood, rarely of metal, could be easily carried in a pocket. They are well worth attention from scientists and collectors (Kisch, 1960b). In Germany, special guilds of craftsmen assumed the right to make the little wooden boxes (*Laden*) in which the money weights were kept, each in its separate recess, each recess inscribed with the name of its coin. These wooden boxes made in Cologne in the sixteenth and early seventeenth centuries were little masterpieces of Renaissance art (Figs. 90, 91). Free-lance craftsmen gradually encroached upon this illegally assumed monopoly of the guild of carpenters.

These boxes are valuable in the study of cultural history because they contain in handwriting or on printed labels the name and address of the maker of the scale and weights and the year of its appearance; all this was prescribed by law (Kisch, 1960b). They also show (rarely) the mastersign of the balance maker and in the seventeenth century also that of the wooden-box maker (Fig. 92).

In collecting coin weights one has to decide on a basic principle for arranging them—according to the maker or according to the coin—because their number is enormous (the Streeter Collection possesses more than a thousand different specimens). In every country where money scales were made, coin weights were of course also made. It therefore would seem logical to arrange a collection of coin weights according to the country of origin, and within the country to subdivide them according to their makers. On the other hand, coin weights precisely follow a country's numismatic changes, which is why Dieudonné and others prefer to list coin weights according to the countries for whose coinage they are valid.

The typical weight and scale boxes for the use of merchants, money changers, jewelers, and pharmacists contained in the lid of the box or in a little compartment within the box the very small grain weights in the form of squares of sheet metal. Sometimes a little instrument for handling the weights was also included, or a nail-shaped object for removing them from the grooves in which they lay. This nail also served as a kind of safety lock for the lid of the grain-weight compartment or for an additional drawer built into the body of the box containing extra coin weights. The small, convenient box, fitting into the pocket of the owner, could thus contain forty or more coin weights in addition to the small scale.

The weight and scale boxes for jewelers were somewhat different; the weights were not equivalent to certain coins but to a certain number of carats. Their scales usually still have deep spherical pans, and as a rule they contain a forceps to handle the weights. This forceps is shaped at the end like a tiny shovel, probably for easier handling of precious stones or pearls (Fig. 93 and p. 55).

The mastersign of the maker is often impressed on each coin weight and on the pan of the scale. This gives the collector or buyer a certain guide to

137

whether the box and its contents are a genuine set. Adjuster marks can also be seen occasionally on the weights and scales. In Nuremberg the mastersign of its maker was impressed on one pan of the scale and the coat of arms of Nuremberg on the other. Figure 94 shows the triangular pan of a coin scale (see p. 69) with the mastersign of the Cologne scale maker Tönnis von Aachen (seventeenth century). The three crowns above his mastersign indicate that at the time of checking this scale he was the sworn Aichmeister of the Free City of Cologne.

Fig. 92. Part of a Cologne weight and scale box showing the sign of the box maker twice.

Fig. 93. Typical Dutch weight and scale box for jewelers (carat scale).

Fig. 94. Triangular pan of a coin scale; seventeenth century. Mastersign of the Cologne balance maker, Tönnis von Aachen, who died in 1655. The three crowns above the initials show that he was at that time the sworn adjuster of the city of Cologne.

139

PHARMACEUTICAL WEIGHTS

In olden times the physician himself prepared and dispensed the medicines he prescribed, and weights and scales were therefore part of his inventory. Probably the oldest pictorial proof of this fact is found on a relief from Kom Ombo in Upper Egypt, supposed to date from Ptolemaic times, which depicts all the instruments of a physician, including a small balance. A picture of this remarkable relief can be found in Laignel-Lavastine (1936, vol. 1, p. 110) and in Thorwald's book (1962, p. 78). The presence of a balance among surgical tools seems to have been accepted as a matter of course by both of these authors.

The distribution of potent drugs has always been a matter of great importance. They had to be weighed carefully, and because many of the prescriptions used through the centuries stem from Greek and Roman tradition, the medical weights faithfully represent the continuity of Roman metrology. In the Middle Ages and later, when everywhere in Europe the merchant's pound equaled 16 ounces, the physician alone retained the traditional Roman 12-ounce pound. The famous scientist and physician Georg Agricola, in his standard work in 1530, warned against the dangerous mistake that occasionally occured, i.e. that the physician or pharmacist used the 16-ounce instead

Fig. 95. Illustration from a manuscript of Dioscorides; Italy, about 1400. It has a short enumeration of all medical weights and their fractions, beginning with the line "opportet sic nosce[re] omnia pondera medicinalia."

of the 12-ounce pound. Medical texts have always stressed the importance of knowing precisely the weight units used in medicine—their names, subdivisions, and symbols. A good illustration of this is Figure 95 from a fourteenth-century manuscript of Dioscorides, which is in the Yale Historical Medical Library. It begins with the definition of the medical weights as a most important piece of knowledge for each physician.

Medicine has not only preserved Roman metrology but has also kept its own signatures and names for its weights (Galen, 1530, 1545). Unfortunately, specimens of small pharmaceutical weights used in the Roman period have not survived. The smallest Roman weights known, preserved in the Streeter Collection and elsewhere, are truncated spheres made from stone, or little lead plates, generally quadrangular. We can only assume that they were used by physicians as well as by goldsmiths and jewelers. The smallest weights of early medieval and Byzantine times, found in Egypt, are similar, but pharmacists later used weights different in appearance from those of the common market. Medical weights in the form of the typical signatures for an ounce, a dram, a scruple, and so on were apparently used in Holland during the eighteenth century and can be seen in Figure 84; it is hardly possible to mistake one of these weights for another. Weights like this were also used in the seventeenth century in Italy. Specimens from Milan with the typical marks showing St. Ambrose and the year of issue are in the author's collection and elsewhere. Similar weights of brass, on which the number of "gross" is indicated by the number of punched holes, were used in the eighteenth century in Germany (Berlin, Nuremberg), Austria, and elsewhere (Fig. 84).

In the countries that accepted the metric-decimal system during the nineteenth century, pharmaceutical metrology was made to correspond to that of the new system, and thus the old denominations and the figurative weights disappeared. As with goods of all other kinds, drugs now by legal decree had to be prescribed and dispensed in terms of grams, decigrams, centigrams, and milligrams. If larger amounts were used, decagrams, hectograms, and so on were used. In the countries that continued to adhere to the time-honored system of ounce, dram, and scruple, pharmaceutical weights have remained up to the present time distinct in form and inscription from other weights.

In the eighteenth century, pharmaceutical weights in the form of a truncated four-sided pyramid were used in Bavaria. On the larger of the two quadrangular planes, the engraved inscription indicated the weight, usually one of the following denominations:

$\tilde{\mathfrak{z}}$	one ounce
$\tilde{\mathfrak{z}}$S or $\tilde{\mathfrak{z}}\beta$	$\frac{1}{2}$ ounce. S or β always stands for the Latin *semis* $(= \frac{1}{2})$

ℨij or ℨii	2 drams (drachms)
ℨj or ℨi	1 dram
ℨS or ℨβ	½ dram
∋ij or ∋ii	2 scruples
∋j or ∋i	1 scruple
∋S or ∋β	½ scruple

These sets were commonly preserved in closed wooden boxes (Fig. 96). The Streeter Collection has a boxed set of pharmaceutical weights and small scale with cylindrical pans similar to the Dutch corn scales (Fig. 22), but its markings prove that it had been made in the eighteenth century for pharmacists. A similar, attractive wooden box with a set of Cologne pharmaceutical weights and small scale is preserved in the Cologne Stadtmuseum in Zeughaus. It also dates from the eighteenth century (Kisch, 1960b). Figures 96 and 97 show typical eighteenth-century examples from Bavaria and from Holland.

In the nineteenth century coin-shaped pharmaceutical weights became more and more popular. In England they bore the royal crown and the inscription "Apothecaries weight." Their denomination was given either in numbers or in the traditional signatures. In the United States, the crown of course was not used, but the denominations were indicated by the familiar medical signs. The maker, who usually was also a scale maker (or the distributor), put his name and trademark and even his business address as an advertisement on the weights. During the coin shortage of the Civil War in the United States, many druggists, businessmen, and weight and scale makers issued tokens bearing their names and addresses.

The names and signs of pharmaceutical weight makers in England and the United States in the nineteenth century can be listed briefly. There are doubtless others, but they have not been encountered in twenty years of collecting.

In England we find coin-shaped weights with the inscription APOTHE-CARIES WEIGHT/ [crown]/ March 16, 1847/ W.& T.A. The initials stand for W. & T. Avery, a well-known firm in Birmingham; 1847 may indicate the date of patent. Other factories apparently also made pharmaceutical weights, for some are marked only with the English crown and APOTHE-CARIES WEIGHT, and others show the crown alone with the word STANDARD around it. Another pharmaceutical coin weight (2 scruples), probably of English origin, bears the maker's initials B. &'S.

In the United States John P. Gruber, a New York scale and weight maker, issued different "one-dram" monetiform pharmaceutical weights in 1863. On one side of this cent-sized copper token is inscribed: APOTH. WEIGHT / [the American eagle on two laurel branches] / ONE DRAM / 1863; on the reverse: JOHN P. GRUBER / [a scale] / New York. Similar Civil War token

142

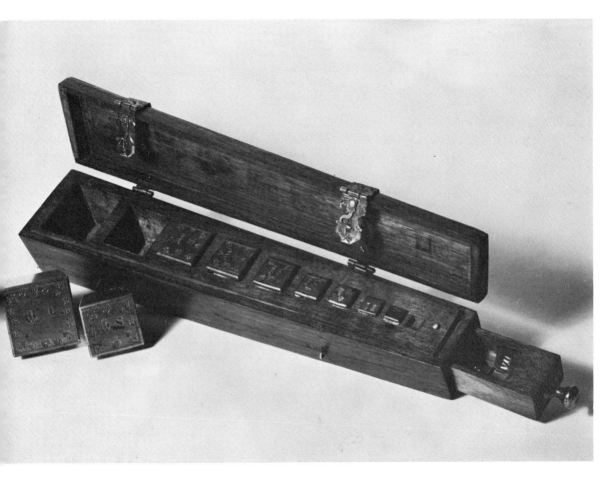

Fig. 96. Typical Bavarian (probably Nuremberg) container of pharmaceutical weights; eighteenth century. The weights are inscribed: 1 pound; $\frac{1}{2}$ pound; 3 ounces; 2 ounces; 1 ounce; $\frac{1}{2}$ ounce; 2 drams; 1 dram. The small weights are in the drawer.

Fig. 97. Box with medicinal weights; Holland, eighteenth century. The large weight is inscribed: LIBR MEDICINALIS.

weights by Gruber have on the reverse the familiar slogan NOT / ONE / CENT, or even the Indian head over the year 1863 surrounded by thirteen stars. Gruber also issued similar weights not meant for apothecaries—for example, one on which a more elaborate scale appears under his name, which also carries his address: 178 Chatham SQ. The reverse shows the inscription GOLDWEIGHT / [eagle on thunderbolt] / TROY / 2 PENNYWEIGHT/ NEW YORK.

The obverse of another Civil War token used as a 1-dram weight is like the Gruber token. The reverse is inscribed: ESTABLISHED / A. [house] D./1850. The house is a two-story, three-windowed structure with the inscription. WARMKESSEL and beneath the house the name of the maker: HORTER. The token is the size of a cent and is of light-colored brass.

Similar weights with pharmaceutical weight symbols were issued by P. ROGERS & CO, and by two firms signing the tokens J.L.B. and W.P.H. respectively. (See also the 20-dollar weight in Fig. 34 with the initials L.C.V.)

Pharmaceutical monetiform weights issued by the firm of H. TROEMNER in Philadelphia are signed with the maker's full name. According to Griffen-hagen (1957), Troemner was a German locksmith who settled in Philadelphia in 1838 and began to make pharmaceutical balances in 1840. Some of his coin weights are signed not with the usual H. TROEMNER PHILA., but only H.T.PHILA., or only H.T. Philadelphia is occasionally abbreviated by the unusual PHILADA.

There are many extant American pharmaceutical weights (drams and scruples) made by John M. Maris; as a rule these are signed only with M in a lozenge: ⟨M⟩. The address is PHILA & NEW YORK.

Objections to the use of monetiform pharmaceutical weights have been made repeatedly, emphasizing that dust accumulates in the deep inscriptions. Gradually they have been replaced in Anglo-Saxon countries by square brass plates with their weight inscribed in low relief or engraved in pharmaceutical symbols. The grain weights are usually squares made of thin sheet metal, the number of grains indicated by impressed dots.

CARAT WEIGHTS

Mention should be made in conclusion of a special kind of weight still used by jewelers for precious stones. The word carat (also *karate, kirat*, in Portugal; *quilada, quilat*, in Brazil) is derived from the seeds of St. John's bread—the carob bean (*Ceratonia siliqua*; Greek, *keration*). As a weight for jewels the carat, according to international agreement, at present equals 200 milligrams, but in Great Britain 205 milligrams. The carat unit of 200

145

milligrams, first suggested by Kunz in 1893, was adopted in France by law in 1909 and in the United States in 1913 (Kunz, 1913).

A carat weight is usually a truncated, four-sided pyramid with the weight inscribed on the base. The entire set of weights, in a small wooden box which also generally contains the scales, consists of individual units with the inscription 1, 2, 3, 4, 8, 16, 32, 64 [carats] (Fig. 93).

INSCRIPTIONS AND ORNAMENTATION ON WEIGHTS

Weights have been more or less adorned according to the aesthetic desires of their producers and users, and their form is an expression of the intellectual and artistic standard of a people. But of still greater importance in the study of a culture are the inscriptions on weights, the basic purpose of which concerns their practical use.

In earlier civilizations it was necessary for the protection of the buyer that he be able to recognize easily the weight represented by a certain piece of metal or stone; second, that he see the government's guarantee of the correctness of the weight; third, that at a time of varying weight standards, the domestic weights be easily distinguishable from the foreign ones. These requirements in large measure explain the different inscriptions that are found on ancient as well as modern weights. Decoration and information were combined in all cases where weights bore the arms of a city or country, and the customer was aware at a glance whether a merchant was using the correct one.

A decorative sign was not always an actual coat of arms. Often, as in Greece, it merely represented an object accepted by a certain community as its symbol and used on its coins. These symbols were the ancestors of the coats of arms which became very popular in the later medieval period.

The following emblems have been found on Greek weights. In Athens, according to Pernice: the astragalus (knucklebone), amphora, turtle, shield, dolphin, half-moon and, in single instances, the cricket or the head of a steer. The entire turtle (say) appeared on a one-*mina* (*mna*) weight; half a turtle on a half-mina weight; and a quarter-turtle on a quarter-mina. This primitive method made clear to illiterate customers the weight of the square pieces of lead or (rarely) brass. The lengthy inscriptions on Babylonian and Assyrian weights were certainly not meant for the man in the street, whereas even the most untutored mind could easily grasp the meaning of one half or one quarter of a turtle on a weight.

The inscriptions on weights have different purposes. On Babylonian ones they indicate the ruler, under whose government or in whose presence the weight had been verified, and the amount of the weight. Egyptian weights

146

may have elaborate hieroglyphic inscriptions (Fig. 43),* but in most cases only the units of weight represented by vertical lines ($|\,|\,|$) are indicated, as they are also on certain Assyrian and Jewish weights. Most commonly used are the following. The figure 10 on all Egyptian weights is indicated by ⌒, which is repeated for multiples; for example ⌒⌒⌒⌒ = 40. The sign O stands for *deben*, also an Egyptian unit of weight, and the gold deben is represented by ⚲ because ⚲ stands for *nub* (gold); O (as a unit of weight in the Old and Middle Kingdoms) is an abbreviation of the hieroglyphs ⫽O and ⫽▪, both standing for deben (Weigall, 1908).

Weigall mentions among the inscriptions on weights in the Cairo Museum one meaning six gold deben: ⚲⫶⫶ and one representing 19 deben: ⌒O⫶⫶⫶ . In the New Kingdom, at the time of the Thirteenth–Eighteenth Dynasties, the *kedet*, a standard unit of weight, was introduced (also called *kite* or *kat* by some Egyptologists). Its symbol is ◀▪ and its weight is 8.8 to 10 grams (Weigall, p. xiv). Up to the time of the kedet, Egyptian stone weights had been oblong or rectangular. Only with the Eighteenth Dynasty did the domed circular stone weight (the cupcake form) come into use. The standard at that time was 1 deben = 10 kedets or 12 "pieces" (⬡ ▦) (Weigall, 108, p. xi). The owner's name was sometimes added to the numerical value of the stone weights. A beautiful example of this is the ovoid weight from the Turin Museum (Fig. 43) which Weigall also mentions. He reads its inscription: "The lector etc. Hepy: 10 Deben." The figure shows the hieroglyphic inscription clearly; it is not quite correctly reproduced by Weigall or by the catalogue of the Turin collection. Egyptian weights have also been recovered that bear the name of the artisan and the name of the commodity (e.g. fish) to be weighed (Glanville, p. 892).

Inscriptions indicating their owner or maker are also known on weights from the classical Roman period.

GREEK WEIGHTS

On Greek weights, the numbers of units are indicated by their names, usually cast in relief (not engraved as in Egypt); sometimes the name of the maker or owner is also found. Weights and scales in Greece were part of the temple inventory, probably for weighing the gold and silver donations of pious visitors. The sanctity of these instruments is expressed in the inscription

---

* I owe this picture to the kindness of the director of the Museo Egizio in Turin, Professor Scarmuzzi.

ΔΙΟΣ (*dios*, of the god) or ΔΙΟΣ ΙΕΡΟΝ (*dios hieron*, consecrated to the god) found on certain bronze weights from Olympia (Pernice, 1894, p. 9). The officially checked standard weights were also kept in the temples or in other generally accessible places. They were made of brass, whereas the market weights used for daily commerce were usually of lead. Some of these weights were standardized and bear an embossed seal, and some carry the added word ΜΕΤΡΟΝΟΜΩ (*metronomon*, by the sealer) which certified that they were officially adjusted by the weighmaster. One can judge from the well-preserved bronze weights of the Greek era that their exactness, whether sealed or not, was far removed from modern standards for market weights, to say nothing of analytical weights.

The most important inscription on Greek weights (indicating their value) in many cases is not too legible, even if time, oxidation, decay, and damage have not affected them. The name of the weight unit was often long and the

TABLE 3.   Denomination of Greek Weights and Gram Equivalents According to Pernice

I. Standard: the heavy Attic unit, the *stater*

		Grams
ΣΤΑΤΗΡ	(stater)	873.2
ΤΡΙΤΗΜΟΡΙΟΝ	(tritemorion)	291.07
ΗΜΙΤΡΙΤΟΝ	(hemitriton)	145.53
ΗΜΙΣΥΗΜΙΤΡΙΤΟΝ	(hemisyhemitriton)	72.77
ΤΕΤΑΡΤΗΜΟΡΙΟΝ	(tetartemorion)	218.3
ΗΜΙΤΕΤΑΡΤΟΝ	(hemitetarton)	109.15
ΗΜΙΣΥΗΜΙΤΕΤΑΡΤΟΝ	(hemisyhemitetarton)	54.57

II. Standard: the light Attic unit, the *mna* (*mina*)

ΜΝΑ	(mna)	436.6
ΤΕΤΑΡΤΟΝ	(tetarton)	109.15
῾ΕΚΤΗΜΟΡΙΟΝ	(hektemorion)	72.77
ΟΓΔΟΟΝ	(ogdoon)	54.57
ΔΡΑΧΜΗ	(drachme)	4.37
ΔΙΔΡΑΧΜΟΝ	(didrachmon)	8.73
ΤΡΙΔΡΑΧΜΟΝ	(tridrachmon)	13.10
ΤΕΤΡΑΔΡΑΧΜΟΝ	(tetradrachmon)	17.46
ΠΕΝΤΑΔΡΑΧΜΟΝ	(pentadrachmon)	21.83

Table 3—continued

Greek weights and their signatures according to Nissen*

			Drachms
		ταλαντον	6,000
T	talanton		
		πεντακισχιλιαι	
⊠	pentakischiliai		5,000
		χιλιαι	
X	chiliai		1,000
		πεντακόσιαι	
⊓	pentakosiai		500
		ἑκατον	
H	hekaton		100
		πεντηκοντα	
Ͷ	pentekonta		50
		δεκα	
Δ	deka		10
		πεντε	
Γ	pente		5
⊢			1
			Obolos
I			1
		ἥμισυ	
C	hemisy		$\frac{1}{2}$
		τεταρτημόριον	
T	tetartemorion		$\frac{1}{4}$
		χαλκοῦς	
X	chalkus		$\frac{1}{8}$

* Drachme means a handful in Greek. It has 6 oboloi, which means as many oboloi (spears of iron) as can be held in one hand. In medical metrology a gramma or scriptulum or scripulum was equal to 2 oboloi; $\frac{2}{3}$ of an obolos was a thermos or lupinus and $\frac{1}{3}$ of an obolos a keration or siliqua.

space available for the inscription small. Abbreviations were therefore common; for instance (Pernice, p. 9) ΣΤΑΤ (*stat* = *stater*). The legibility is not enhanced by the fact that such inscriptions were not always written in one line but were arbitrarily divided to fit the space. Thus the four letters on the weight of a stater were distributed in its four corners as follows ⊠ (Λ was commonly used for capital alpha).

149

Table 3 gives the names and supposed values for the weights used in ancient Greece, according to Pernice (p. 49), and weights and their signatures according to Nissen (p. 869). Stater was not only the name for the double drachme (*didrachmon*) but also for the double mna. We shall refrain here from entering the field of Greek metrology, which is a subject in itself and an interesting and difficult branch of archaeology, and confine ourselves to the instruments for weighing. The table is presented only for the sake of the inscriptions on the weights.

Roman and Byzantine Weights

Probably the most important publication on Roman weights is the monograph by Karl Pink (1938). It is regrettable that the book is so brief and the catalogue confined to objects in Austria, but in most scholarly fashion the author brings together in the introduction the main facts pertaining to this topic.

Very few early specimens of Roman weights with inscriptions naming the maker or the owner are extant. In a few instances the place where standard weights were kept is indicated. This is the case with weights from the temple of Jupiter Capitolinus in Rome where, in the second century B.C., the office of the *aedils* who were charged with the supervision of weights and measures was still located (Pink, p. 51). The temple of Castor in Rome also housed standard weights and various sets exist which are inscribed with the amount and the word Castor or abbreviations of it such as CASTO, CAST, CAS, CA. In imperial times the temple of Castor was a fiscal depot and a sealer's office for adjusting weights (Pink, p. 53). The Roman army used a special type of weights in foreign areas which bear the inscription CASTROR AUG or the name of the regiment to which they belonged (for example, *legionis primae Italic*). Pink's monograph gives an instructive account of inscriptions on Roman weights.

Roman and the later Byzantine weights carry symbols for certain words like *nomisma* or *solidus* which must therefore have been well known and understood by the common man (Fig. 87). In Roman literature, especially in scriptural passages on human and veterinarian medicine, one also finds the usual symbols for weights, measures, and numerals. A few tables covering this field (Paucton, 1780) have therefore been reproduced.

Up to the third century A.D. stone weights were most common in Rome, but some of bronze and occasionally even of lead have been found in Pompeii. Drawing on Pink's monograph and wide personal experience, several tables have been constructed.

TABLE 4.   Subdivisions and Symbols of the Roman Pound in the
First Century B.C.*

Name	Symbol	Name	Symbol
As	I	Sescuncia	Ⅽ, Σ·
Deunx	S::.	Uncia	., —, o, ∪ ∼
Dextans	S::	Semuncia	Ⅽ, Σ, E
Dodrans	S:.	Duella	Π
Bes	S:	Sicilicus	Ɔ
Septunx	S.	Sextula	Ϩ
Semis	S	Dim. Sextula	Ϩ
Quincunx	::.	Scripulum	Ɔ, ϶
Triens	::		
Quadrans	:.		
Sextans	:		

* *As* = pound = 12 ounces; quadrans = ¼ *as* = 3 ounces; sextans = 1/6 *as* = 2 ounces.

Symbols for numbers of pounds on ancient Roman weights were C for 100 and ↓ or ↑ or ⊥ or L for 50. Eighty-five pounds, for example, was written ↓XXXV. Table 4 shows the subdivisions of the pound and their symbols in the first century B.C. (Pink, p. 24). Appendix 1 gives the value of these units and their weight in grams.

The number of pounds was also indicated on the weights by Roman numerals like I, II, III. Sometimes the figures were marked in dotted lines: ⸭

(= X), like the X on a ten-pounder of serpentine in the Streeter Collection, where one of the branches of the X is lengthened by one extra dot, thus changing the X to a Latin cross (Kisch, 1959). In this early period the number of ounces was always indicated on stone weights by dots or small circles; for example, the two 3-ounce weights of serpentine in the author's collection are each denominated thus: ∴. A 6-ounce weight is usually marked S, the initial standing for the Latin word *semis* (half), because 6 ounces equal one half a Roman pound. The symbol S has survived to the present on pharmaceutical weights where, for instance, ꝫS means ½ dram. However, a 6-ounce weight of bronze (fourth or fifth century) in the author's collection is marked by six dots in the following arrangement: ⸭⸭ . Sometimes the early Roman weights have a P after the numeral, indicating *pondo* (pound). The inscription VNCIA has also been found. One and one half is occasionally expressed SSS (i.e. $3 \times \frac{1}{2}$) or 150 = LLL (i.e. 3×50).

151

TABLE 5.   Weight Symbols Used in the Late Imperial Roman Era

1 Pound	Λ, ΛΛ, Λ Λ
Ounces	
6	⅄ Ϛ
5	⅄ E
4	⅄ Δ
3	⅄ Γ
2	⅄ B
1	⅄ A
½	IB (12 scruples)
¼	Ϛ (6 scruples)

TABLE 6.   Numerals on Byzantine Weights

Pounds*	
3	Λ Γ
2	Λ B
1	Λ Λ
½	↑Λ , ⌐° S (6 ounces)
Ounces	
3	⌐° Γ
2	⌐° B
1	⌐° A
½	IB, XII
	(12 scruples)

* For Λ (meaning libra = pound) this writer has occasionally also found the symbol Λ.

The symbol for uncia (the sign ⅄ , which is transcribed in manuscripts as ᶚ or ⅄) appears on Roman weights at the end of the fourth century (straight lines are less difficult for the engraver than curved ones). A symbol of tantalizing similarity has been found on various Hebrew weights excavated in Lachisch: Ⴟ or ⵝ (88.7 to 92.4 gm). This similarity of symbols merits further metrological investigation (Tufnell, 1953, p. 353 and Pl. 51; Diringer, 1934, p. 263; Yadin, 1960).

The words *litra* and *libra* for pound, later abbreviated to lb, ℔ appears as a Greek lambda (Λ) on weights, especially in the later Byzantine period. A lambda with a subscribed iota Λ or Λ, rarely Λ, was also used.

152

TABLE 7. Symbols for Nomisma on Byzantine Weights

72 = N OB	10 = NI	
40 = N M	9 = NΘ, ✧	
36 = N ΛS	8 = NH	
30 = N Λ	7 = NZ	
26 = N KS	6 = NS	
25 = N KE	5 = NE	
24 = N KΔ	4 = NΔ	
20 = N K	3 = NΓ	
18 = NIH	2 = NB	
17 = NIZ	1 = NA, N	
15 = NIE	$\frac{1}{2}$ = IB (12 siliquae)	
12 = NIB	$\frac{1}{3}$ = H (8 siliquae)	

* N stands for nomisma. One also finds ∇ for 10 drachms and ∇ ∇ for 20 drachms.

Greek numerals were used in Rome to indicate the number of weight units. In the following Byzantine epoch the Greek influence was still more powerful, and the Greek form of square plates for weights, bearing the Greek numerals, was adopted.

The symbols in the late imperial period are shown in Table 5. Symbols on the square weights sometimes differed from those on the disk, and new denominations occasionally appear (Table 6). In A.D. 307, when Constantine I introduced into Roman coinage the solidus, a gold coin weighing $\frac{1}{72}$ pound, it became, under the name *nomisma*, a standard unit for market weights in the middle of the fourth century. This coin stabilized Roman currency (Pink, p. 33). From that time the symbols on the weights for pound (Λ, Λ), for ounce (ϝ), and nomisma (Ν̂) remained in use for many centuries (Table 7).

A horizontal line was also used for the ounce (Fig. 87). Pink (p. 92) depicts a sixth-century Byzantine weight with the inscription for 1 ounce (6 solidi) in three lines as —I/SOL/TϹT. Byzantine-type weights from the sixth century, whose place of origin according to Pink (p. 40) was Africa, are somewhat different in appearance. They are inscribed as solidi, not nomisma. According to Pinkerton (1789, vol. 1, p. 156) the Byzantine solidi (weighing $\frac{1}{6}$ ounce) in Europe were also called *bezants*; Byzantine writers called the nomisma (the coin), *crysinos* (the golden one) or *hyperperos* (shining like fire). Sometimes the Byzantine weights are inscribed S or Sol.*

* The unreliability of Pinkerton's statements, for instance, concerning old Jewish coins, is proven by the following: Henry Lawrence wrote on November 28, 1858, in the Yale library copy of Pinkerton's book that he had met Pinkerton between 1808 and 1812. Pinkerton told him that he was not conversant with the subject of numismatics and that the materials for the book had been given to him only for the purpose of compilation.

A horizontal line at the level of the top of subsequent numerals always indicates uncia. For example, on the ⅓-pound-weight (= 4 ounces or 24 solidi): — | | | | or SOL XXIIII; on a 1-ounce weight (= 6 solidi): —| or Sol Ϛ. Similar inscriptions of this period also used, instead of the horizontal bar, the Byzantine symbol for uncia Ϝ , and that for pound Λ (Fig. 87). Even the identical denominations nomisma and solidus occasionally appear next to each other on the same weight; for example, NΛ (Λ = 30) and SOL XXX (Pink, p. 41). We also find in inscriptions on some weights of this time VS, VSL, or VSVALE, abbreviations of *moneta usualis* (current money) and also X or DN, both standing for *denarius*, a coin equal in weight to one nomisma (Pink, p. 42; Sabatier, Pl. II).

The inscriptions on late imperial and Byzantine weights were engraved and usually inlaid with silver. Often this silver intarsia was lost in centuries of use or maltreatment. Sometimes only N is still legible instead of Ν̇, and Γ instead of Ϝ. Greek letters continue to be used for numerals; for example I, K, L, M, etc., designating 10, 20, 30, 40, etc.

Inscriptions on Byzantine weights that give scholars in this field many headaches are the monograms. They might reveal much, if one were only able to decipher these artistic, often very appealing, combinations of letters. Monograms on Byzantine coins were common in the fifth and sixth centuries A.D. and are also found on weights from the Byzantine period, but their solution is usually restricted to wishful thinking.

Engravings and adornments were used increasingly during the Byzantine period to embellish weights. Busts of emperors, like those on coins, are seen on some of them. This custom was still followed up to the eighteenth century. On the heavy French market weights it is not uncommon to find a bust of the king. One of 25 pounds with the portrait of Louis XVI, for example, is in the Streeter Collection. Money weights with portraits of rulers are not rare up to the early nineteenth century.

The Roman exagia and the Byzantine weights were apparently the first to display portraits of emperors (see Fig. 87). Saints were also depicted (Pink, p. 66), and branches and garlands were often used as ornaments. On the typical Byzantine weights, if they have any inscriptions at all, the sign of the cross is rarely absent. One rare exception is a 6-ounce (Γ S) weight (No. 37306), preserved in the collection of the Bezallel Museum in Jerusalem, on which there is a six-pointed star instead of a cross. If the cross is used, it may be in the form of a T (tau cross), or with four equal branches + (Greek cross), or with unequal vertical branches †(Latin cross). Pink also mentions as a Christian symbol on weights, the chrysmon ⳨, and emphasizes (p. 29) its similarity to the symbols ⵣ and × for the *denar*. It has also been used as a symbol for nomisma on weights of the Byzantine period. The symbol A–Ω has also been

154

found, and the author has seen engraved columns as ornaments on Byzantine weights, as well as a six-branched star and birds-eye (**ⵔ**) ornamentation. Rarely, one finds Christian benedictions engraved on Byzantine weights, like the very short one ΘΕΟΥ ΧΑΡΙC (*Dei gratia*: Pink, pp. 61 ff.), sometimes contracted to a kind of monogram ⴱⴼⵁ.

FRENCH WEIGHTS

The Roman influence on the shape of weights in France was long noticeable. In medieval ages and up to the eighteenth century they were disks or square plates. The pound and its fractions (*miei, mezo* = $\frac{1}{2}$; *cartero, cartaro* = $\frac{1}{4}$; *onza, uncia*) were in most cases the inscriptions used, plus the year of issue. This latter fact probably indicates that French weights at this time (as with the coins) were issued by the government in limited quantities. However, in the sixteenth to the eighteenth centuries private (but licensed) weight makers in Antwerp and Amsterdam also put the year of issue on money weights. This may have been necessary to indicate the year in which they were legal coin weights, because the weight of gold coins was changed frequently by the governments.

In ancient France the cities issued weights according to the local standards. Even in the same town the governor occasionally had a standard for his castle that was different from that accepted in the town proper. Nearly every issue of medieval French city weights bears either the coat of arms of the town or a well-known landmark such as a famous church (e.g. Toulouse, Fig. 55), apparently to make the provenance of a weight and its standard easily recognizable in the market to the illiterate, as the majority of customers at that time were.

Table 8 records the French cities from which specific weights have survived. The data and symbols are compiled from Gaillardie, Machabey, Forien de

TABLE 8.   Known Weights from French Cities

The following symbols are used:

○ = round disk	8 = octagonal disk
6 = hexagonal disk	5 = pentagonal disk
□ = quadrangel	TrPy = truncated pyramid
⚜ = fleur-de-lis	℔ = pound

The figures in bold type indicate that the only known specimen is in the Streeter Collection.

Table 8 continued.

City	Form	Denomination	Symbol	Date of Issue
Agde	8	**1**, $\frac{1}{2}$, $\frac{1}{4}$, ℔	shield with 3 or 4 wavy lines	
Albi	○	2, 1, $\frac{1}{2}$, $\frac{1}{4}$, $\frac{1}{8}$, ℔ 1 ounce	*B*, tower with or without leopard passant	1336, 1581, 1493, 1583, 1517, 1675, 1557
Alet	○	$\frac{1}{8}$ ℔ 1 ounce	cross, crosier	
Ardres	TrPy 6	$\frac{1}{2}$, $\frac{1}{4}$, ℔ 1 ounce	double eagle and 3 ⚜	
Arles	○, 8, 6, ☐	1, $\frac{1}{2}$, $\frac{1}{4}$, $\frac{1}{8}$, ℔	sitting, rarely passant, lion	
Arras	Bell shape	$\frac{1}{2}$ ℔ 2 ounces	Gothic letter "a"	
Auch	○, 8	4, 2, $\frac{1}{2}$ ℔	rampant lion, crosier, lamb and cross	1309
Aurillac	○, Truncated cone	2, $\frac{1}{2}$, $\frac{1}{4}$, ℔	3 ⚜ and 3 shells in shield	
Auvergne	8	$\frac{1}{4}$, $\frac{1}{8}$, ℔	arms	
Avignon	○, 8, TrPy	2, 1, $\frac{1}{2}$, $\frac{1}{4}$, $\frac{1}{8}$, ℔ 8, 4 ounces	a single or 2 crossed keys	
Bayonne	○, Cylinder with handle	100, 50, 25, 10, 5, 4, 3, 2, 1, $\frac{1}{2}$, $\frac{1}{8}$, ℔	tower, or 3 ⚜	1529
Beaucaire	8	1 ℔	arms under ⚜	
Béziers	○, 8, TrPy 8	2, 1, $\frac{1}{2}$, $\frac{1}{4}$, $\frac{1}{8}$, ℔ 1 ounce	3 straight lines under 3 ⚜	1578
Bordeaux	○, Bell shape	2, 1, $\frac{1}{2}$, $\frac{1}{4}$, ℔	passant lion, tower with open door	1316
Bretagne	○, TrPy	50, 25, $\frac{1}{2}$ ℔	head ⚜	

City	Form	Denomination	Symbol	Date of issue
Cahors	O	$\frac{1}{2}, \frac{1}{4}, \frac{1}{8}$, ℔	3 towers, bridge	
Carcassonne	O, TrPy 8, 6, 5	2, 1, $\frac{1}{2}, \frac{1}{4}, \frac{1}{8}$, ℔ 1 ounce	lamb with flag, ⚜, 2 towers at gate, monogram CAR	1555 1578 1667 1675 1678 1693
Carpentras	8, 6	6, 4, 3 ℔ 100 gm	⚘	
Castelnaudary	O	4, 2, 1, $\frac{1}{2}, \frac{1}{4}$ ℔	3 towers ⚜, B A	1512 1763 1765
Castelnau-de-Montmirail	O	$\frac{1}{2}$ ℔	3 towers, ⚜	1293
Castelsarrasin	O, Bell shape	2, 1, $\frac{1}{2}, \frac{1}{4}$ ℔	3 towers ⚜, Maltese cross	1274 1580
Castres	O, Bell shape	4, 2, 1, $\frac{1}{2}, \frac{1}{4}, \frac{1}{8}$, ℔ 1, $\frac{1}{2}$ ounce	rampant lion, arms of C., ⚜	1358, 1639 1380 1594
Caussade	O	1, $\frac{1}{2}, \frac{1}{4}$ ℔ 1 ounce	tower, ⚜	1518 1578
Clermont-Oise	O		tower under 3 ⚜	1765
Condom	O	1, $\frac{1}{2}, \frac{1}{4}, \frac{1}{8}$, ℔ 1 ounce	2 parallel keys, 2 towers	1229 1280 **1334** 1368 1466
Corbie	O	$\frac{1}{4}, \frac{1}{8}$ ℔	goose, crosier, 2 keys	
Cordes	O	1, $\frac{1}{2}, \frac{1}{8}$ ℔	wall with 3 towers, ⚜	1280 1293
Dijon	6	4, 2, $\frac{1}{4}, \frac{1}{8}$ ℔	arms	
Fleurance	O	$\frac{1}{4}$ ℔	double eagle, rosette	1245

(*continued on next page*)

157

Table 8 continued.

City	Form	Denomination	Symbol	Date of Issue
Foix	○, 8	$1, \frac{1}{2}, \frac{1}{4}, \frac{1}{8}$ ℔ 1 ounce	shield with vertical stripes	1299
Fougères	8	$\frac{1}{8}$ ℔		
La Française	○	$\frac{1}{2}$ ℔	Maltese cross, ⚜	1578
La Magistère	8	$\frac{1}{2}$ ℔	cross under 3 ⚜	
Gaillac	○	$\frac{1}{2}, \frac{1}{8}$ ℔ $\frac{1}{2}$ ounce	cock, ⚜	1291
Lectoure	○	$2, 1, \frac{1}{8}$ ℔ 1 ounce	bishop, standing lamb, ox	1307 1346 1528
Lézignan	8, 5, □	$\frac{1}{2}, \frac{1}{4}, \frac{1}{8}$ ℔	shield with 3 ⚜	
Lille	○, 8, 6, Rhomboid	$2, 1, \frac{1}{2}, \frac{1}{4}, \frac{1}{8}$, ℔, 2, 1, $\frac{1}{2}$ ounce, 2 deniers, 200, 100 gm	⚜	1734
Limoux	○	$1, \frac{1}{2}, \frac{1}{4}, \frac{1}{8}$ ℔ 1 ounce	St. Martin on horse, ⚜	1270
Marseille	8	$1, \frac{1}{2}$ ℔	cross in shield	
Millau	8, TrPy	$\frac{1}{2}, \frac{1}{4}, \frac{1}{8}$ ℔	shield under 3 ⚜	
Mirepoix	○, 8	$1, \frac{1}{4}, \frac{1}{8}$ ℔ 1 ounce	fish, shield, with 3 oblique lines at each side	1275 1310 1311 1315
Moissac	○	$1, \frac{1}{2}$ ℔	cross under 3 ⚜, 1 ⚜	1572 1573
Montagnac	○	1 ℔	tree on hill, ⚜	
Montauban	○, 8	$1, \frac{1}{2}, \frac{1}{4}$ ℔ 2 ounces	tree on hill, ⚜	1304, 1551 1345, 1568 1347, 1572 1348, 1573 1448? 1578

City	Form	Denomination	Symbol	Date of Issue
Montbard	○		2 fish	1587
Montolieu	○	$\frac{1}{8}$ ℔ 1 ounce	olive tree, crosier	
Montpellier	○, 8	4, 2, 1, $\frac{1}{2}$, $\frac{1}{4}$, $\frac{1}{8}$ ℔ 1, $\frac{1}{2}$ ounce	shield with dot in center, ⚜	1559 1604
Montpezat	○	1, $\frac{1}{2}$ ℔	balance, cross on sphere	1506 1514
Montreal d'Aude	○	1 ℔ 1 ounce	tree on hill, ⚜	"Philipus Rex"
Nancy	○		crown over shield, 3 birds on oblique beam	
Narbonne	○, 5, 8	2, 1, $\frac{1}{2}$, $\frac{1}{4}$, $\frac{1}{8}$, ℔ 1 ounce	in shield, key and double cross under 3 ⚜	1595, 1665 1601, 1666 1644, 1668 1646, 1671 1648, 1674 1652, 1675 1653, 1677 1655, 1678 1657, 1679
Nérac	8		sun	
Nîmes	0, 5, 6, 8	1, $\frac{1}{2}$, $\frac{1}{4}$, $\frac{1}{8}$ ℔ 1, $\frac{1}{2}$ ounce	palm tree with crocodile tied to it, tower	1577 1596 1793
Orange	○, □, 8	1, $\frac{1}{2}$, $\frac{1}{4}$, $\frac{1}{8}$ ℔	trumpet	1689
Orthez	○	1, $\frac{1}{2}$, $\frac{1}{4}$ ℔	cow, gate with 2 towers and 2 keys	1274 1515 1612
Pamiers	○	1, $\frac{1}{2}$, $\frac{1}{4}$, $\frac{1}{8}$ ℔	rider on horse-back, ship, 3 towers	1240 1247
Perpignan	TrPy 8	4, $\frac{1}{2}$, $\frac{1}{8}$ ℔	⚜	

(*continued on next page*)

159

Table 8 continued.

City	Form	Denomination	Symbol	Date of Issue
Pézenas	8, TrPy 8	2, 1, $\frac{1}{2}$ ℔	arms: 3 ⚜ horizontal stripes and fish	1790
Puycelci	○, 8	1, $\frac{1}{2}$ ℔	⚜, gate	
Rabastens	○	2, $\frac{1}{4}$, $\frac{1}{8}$ ℔	root tuber with 6 sprouts	1241 1249
Rodez	○, Bell shape	2, 1, $\frac{1}{2}$, $\frac{1}{4}$, $\frac{1}{8}$ ℔	rampant lion, shield with 3 rings, crosier	1358, 1650 1501, 1670 1520, 1722 1625, 1776
Rouen	Brick form	100, 50 ℔	crown over 3 ⚜	1738 1789
Saint Afrique	○, 5, 8	$\frac{1}{2}$, $\frac{1}{4}$ ℔	cross beneath 3 ⚜, in shield, ⚜	
Saintes Maries de la Mer	□	1, $\frac{1}{2}$, $\frac{1}{4}$, $\frac{1}{8}$ ℔	2 women standing on boat	
Saint Étienne	TrPy	3 ℔	key and 2 branches	
Saint-Nicolas de la Grave	○	1 ℔	⚜	1644
Saint Omer	TrPy 6, ○, 6, 8	2, 1, $\frac{1}{2}$, $\frac{1}{4}$, $\frac{1}{8}$ ℔	double cross	18th century
Saint Pons	○, 8	3, 2, 1, $\frac{1}{2}$, $\frac{1}{4}$ ℔	tree between S and P	
Saint Porquier	○	$\frac{1}{2}$, $\frac{1}{4}$ ℔	swine under tree	
Saint Thibery	8	1, $\frac{1}{2}$, $\frac{1}{8}$ ℔	3 ⚜ and 3 tubers (?)	1791
Salon-de-Provence	8	$\frac{1}{4}$ ℔	in shield 3 ⚜ over rampant lion and ⚜	
Sauveterre	○	1, $\frac{1}{4}$ ℔	cow, cross	1237 1297 1324

City	Form	Denomination	Symbol	Date of Issue
Saverdun	O	1, $\frac{1}{2}$, $\frac{1}{4}$ ℔	monk, wall with 4 pinnacles	1267
Seyne	TrPy 6	1, $\frac{1}{4}$, $\frac{1}{8}$ ℔	arms	**1599, 1627** **1600, 1668*** **1613**
Strasbourg	O, Bottle shape	1, $\frac{1}{2}$, $\frac{1}{4}$ ℔	bishop, church with 3 towers	1249
Tarascon	8	1, $\frac{1}{2}$ ℔	turtle?	1690
Toul	5	$\frac{1}{2}$ ℔	horn	
Toulon	TrPy, 8, Bottle shape	6, 4, $\frac{1}{2}$ ℔	⚜	1600
Toulouse	O	4, 2, 1, $\frac{1}{2}$, $\frac{1}{4}$, $\frac{1}{8}$ ℔ 1, $\frac{1}{2}$, $\frac{1}{4}$ ounce	church with high tower, wall with 3 pinnacles, Maltese cross	1238, 1450 1239, 1495 1259?, 1500 1274, 1504 1443, 1516 **1527**
Troyes	Bell shape	25 ℔	crowned shield, 3 ⚜	
Valencienne	O	2, 1 ounces	⚭	
Villefranche de Rouergue	5, 8	$\frac{1}{4}$ ℔	2 columns, over each a ⚜	
Villeneuve de Rouergue	O	1, $\frac{1}{2}$ ℔	Maltese cross, ⚜	1341

* In the author's collection.

Rochesnard and Lugan, and the Streeter Collection. For identifying such specimens the excellent "album" of Forien and Lugan is indispensable. A tabulation such as this is revealing in different ways. The number of extant specimens of weights of a certain town and especially the frequency of issue give at least an approximate idea of the importance of commerce and of the weight-making industry in these towns in different centuries. It is further

161

striking that the coin-like weight (*poids monetiform*) bearing a long inscription and a symbol is typical of the cities of southern France.

It is of interest (and merits some historical explanation) that weights bearing the year of issue first appear in the thirteenth century. None is known from the twelfth, but during the early thirteenth century no fewer than sixteen different cities in southern France issued weights which have survived.

TABLE 9.   Letters Representing French Mints*

A	Paris
B	Rouen
C†	St. Lô (later Caen)
D	Lyons
E†	Tours
F†	Angers
G†	Poitiers
H	La Rochelle
I	Limoges
K	Bordeaux
L	Bayonne
M	Toulouse
N	Montpellier
O†	Saint Pourçain (later Riom)
P†	Dijon
Q	Châlons (later Narbonne, then Perpignan)
R	Saint André (later Villeneuve-les Avignon. then Orléans)
S†	Reims
T	Sainte Menehoud (later Nantes)
V†	Thurin (later Amiens)
X	Villefranche-de-Rouergue (later Aix-de-Provence, then Aix)
Y†	Bourges
Z†	Dauphiné (later Grenoble)
𝒢†	Bretagne (later Rennes)
AR†	Arras
♛/L	W. Lille-en-Flandre
♛/S†	Troies
AA	Metz
BB	Strasbourg
CC†	Besançon
[a cow]	Pau

* Dating from the ordinance of François I, of January 1549.

† Mint did not operate after 1780.

162

Hallmarks and Sealer's Marks on Weights

In addition to the mark of the maker, further identifications were inscribed on weights. Among them, and of outstanding importance, were the marks embossed by the government-appointed sealer or adjuster, which validated the weights. Sealing of tested market weights by a special officer (the *muhtasib*) was mentioned as early as the fourteenth century in the Islamic literature (Levy, 1938).

In Europe the marks for sealing weights were designated by the government. In France, between 1549 and 1780, letters of the alphabet were used as symbols of the different mints and to mark the adjusted weights. Table 9, taken from Paucton's book (1780, p. 52) lists these letters.

For Cologne, the official and exclusive sign of the adjuster was three crowns (the arms of the city).

The sign for London was a dagger; for Paris and France a fleur-de-lis; for Barcelona, Bern, Basel, and many other cities, their arms. In England, James I empowered the Plumbers Company to assay and mark all weights of lead made or sold by any of their members, and the Founders Company to do the same with weights of brass (Brewer, 1853).

Among the inscriptions on weights, the impressed mastersigns are most instructive and therefore valuable. They were not assigned by the government; each master was at liberty to choose his own, which was then registered, as is done today with the trademark. The only difference was that the guild imposed on its members the obligation to use such a sign. A master had the right to change his sign, and Table 10 (p. 164) shows that this often occurred. This is true, for instance, with the Free City of Cologne, where a master, when appointed adjuster, always changed his previous mastersign so that it contained then the three crowns of Cologne (Fig. 94). If the workshop of a master had a good reputation, his mastersign was a valuable asset. After the death of the original owner it was either transferred to his heirs or even sold. This we know from the very instructive study of Stengel (1915/1918) on the mastersigns of the Nuremberg coppersmiths. They are most important in attributing old nested weights to a certain period or master. It is particularly noticeable that many Nuremberg mastersigns are similar, and this may not have been due to coincidence or a lack of inventiveness but to the desire to choose a sign that could easily be mistaken by the customer for that of a better-known workshop in the same city.

Although mastersigns and sealers' marks are found on every kind of weights and scales, the coin weights in the scale boxes of the money changers and jewelers provide the surest identification of the marks because these weights bear not only the mastermark but also (especially in Belgium and

163

TABLE 10. Known Scale and Weight Boxes Bearing Mastersigns

Sign No. from Chart 4	Number of Boxes Known	Used in Cologne	Used in Holland (A = Amsterdam)	Used in Antwerp	Location Unknown	Date
5	2	2				1590
4	3	2			1	1592–1608
11	7	4			3	1596–1605
36	1	?		?		1598
20	4	4				1606–1645
31	1	1				1611
18	5		5A			1618–1660
17	1		1			1622
30	9		7A[+1*]		1	1622–1637
16	8		6A[+1]		1	1624–1652
24	1	1				1624
32	1		1A			1630
26	9	5[+1]	1	1	1	1630–1652
34	5		5A			1635–1649
33	4		4A			1641–1660
13	6		6A			1641–1660
10	3	2			1	1618–1645
8	15	10[+1]	2	1		1646–1663
22	2		2A			1649–1656
1	1	1				1649
29	2		2A			1650–1658
27	1	1				1651
7	2	1	1			1650–1659
23	1		1A			1659
35	1	1				1662
36	1			1		1598

* Attribution uncertain.

Holland) their date of issue (see Chart 7). They inform us even today of the efficiency of certain workshops and, to a certain degree, of the commerce of that period.

Embossed letters on coin weights from Nuremberg, which are neither sealers' marks nor mastersigns, are the letters AG and RA, which stand for *Abgezogen Gewicht*, and *Recht Abgezogen*. Both these expressions, which were common in the eighteenth century, mean properly adjusted.

In England an embossed crowned C was the sealer's mark under Charles I (1625–49). A statute of Henry VII in 1491 demanded that standard weights and measures (of brass) be marked by a chief officer with a crowned H, which was chosen in 1495 for all weights used by merchants (Brewer, 1853, p. 310).

PART II
Manufacture of Scales and Weights

6. MASTERSIGNS

In medieval times crafts were organized in Europe, especially in France and in Germany, in union-like societies called guilds or brotherhoods (*Gilden, Aemter, Bruderschaften*). Each of them was ruled by a small group of governors elected strictly according to the constitution and by-laws granted to the group by the government. The weight and scale makers of Cologne, where their history is well known (Kisch, 1960b), were organized into a brotherhood which was first included in the guild of the blacksmiths (*Schmiede*). They became an independent guild with their own sworn constitution in 1553. The scale makers (*balanciers*) in Paris were organized in 1325, in Rouen in 1415, and in Lyons probably in 1668 (see the valuable reports of Testut, 1946, pp. 164 ff. and of Machabey, 1949).

The Constitutions of the guilds reveal a great deal about the cultural history of weights and scales from 1300 to 1700. Every master watched carefully and jealously over each paragraph of the constitution of his guild and the rights and privileges it granted, as well as the duties it imposed. Many of the guilds (for example, in Cologne and Nuremberg) had already made it compulsory in medieval times for each master to choose a sign and imprint it on every piece of goods leaving his workshop. This identification was meant to prevent carelessness or cheating. This rule was also obligatory for the weight and scale makers. The importance of ancient mastersigns is easy to understand; when they are deciphered, they permit attribution of an object to a certain time, place, and workshop. I suspect that this medieval law may have been the origin of the custom since the fifteenth century for European artists to sign etchings, woodcuts, and so forth with their initials, monograms, chosen symbols, and later with their entire names. But this was an earlier custom also, for there are extant Roman weights bearing the maker's name, and Pink (1938) supposes that the stone weights of this time were made by the "marmorarii"—the marble workers in Rome—an early form of guild.

In medieval times and later, the great centers of commerce were also the great centers for the manufacture and export of weights and scales, as the relics prove. Among the outstanding German centers were the independent cities of Cologne, Nuremberg and, at the end of the eighteenth and early nineteenth centuries, Berlin and certain cities in the Palatinate like Barmen and Essen. In France it was Lyons and Paris, Antwerp in Belgium, Amsterdam in Holland, Milan and Turin in Italy, London and (later) Birmingham in England, and Geneva in Switzerland.

167

Export of the usual market weights was not feasible because the standard for these weights changed not only from one country to another but also between counties and often even between towns. Export of standard weights from one capital to another like Cologne or Paris was limited to occasions when governments requested exact replicas of standard weights for introduction into their own countries. However, the weight and scale sets in boxes contained coin weights that were equivalent only to the common gold and silver coins which were valid everywhere. Like the balances, they could be exported, and they constituted a valuable industry in the commercial centers.

In addition to the signs of the weight and scale makers, there are the master-signs of the manufacturers of the elaborate wooden boxes of the sixteenth and seventeenth centuries (Figs. 89–92). Less is known about the box-making masters than about the weight makers. The signs were usually those of masters of the carpenters' guilds in Cologne or in Holland (Chart 4, p. 173).

The export of nested weights became a highly developed industry; they were all made in Nuremberg up to the second half of the eighteenth century, and many were exported. They were adjusted by local masters to the standards of the different countries to which they were sent (see p. 126). According to the rules of the guild of coppersmiths in Nuremberg, each nested weight was marked with the sign of the master from whose workshop it came; thus we are able today to prove the origin of Nuremberg nested weights found in various countries. Even if the local dealer or adjuster had tried to efface the original mark, the place where it had originally been embossed is usually recognizable by the eradication mark.

The known mastermarks found on weights, scales, and wooden boxes are listed in the charts and tables that follow; they were collected by this writer over many years. Wherever the works of other scholars have been incorporated into the lists, the sources are mentioned. The lists are still not complete, and additions will probably be made in coming years.

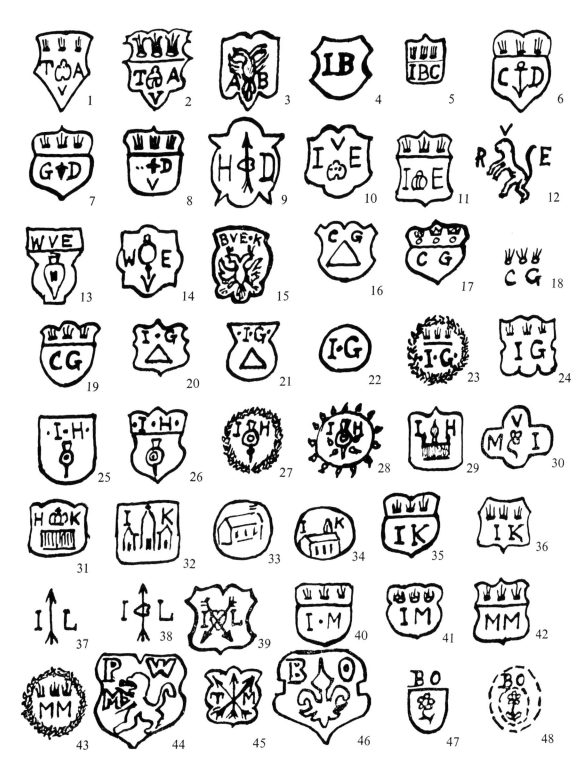

Chart 3. Mastersigns of weight and scale makers in the Cologne guild.
Continued on next page.

COLOGNE MASTERSIGNS

Key to Chart 3. The dates after each name indicate the first and the last year for which the author was able to indentify extant works of this master. A dagger before the year indicates the year of the master's death (proven by archival documents).

Three crowns in a mastersign indicate that a master, at the time it was used, was the sworn adjuster of Cologne.

1, 2.	Tönnies (Antonius) from Aachen, 1648–1655
3.	Arnold from Bochum (Boichem), 1590–1618
4.	Johannes Baum, 1663
5.	Johan Baptist Cöllen (Coellen), 1752–1797
6.	Unknown, 1602
7, 8.	Probably Gerhard from Düssel (Dassel), 1592–1596
9.	Hermann from Düssel (Dassel), 1608–1611
10, 11.	Jan from Essen, 1590–1608
12.	Rutger from Essen, 1590–1615
13, 14.	Wilhelm from Essen, 1624–†1655
15.	Unknown
16–19.	Caspar Graefenberg (Grievenberg), 1703–†1738
20, 21.	Jacob Graefenberg, 1730–1765 (son of Caspar)
22–24.	Johan Graefenberg, 1738–†1753 (b. 1706; son of Caspar)
25–28.	Jacobus Heuscher, 1661–1699
29, 30.	Unknown
31.	Heinrich Kirch (Kirchen), Jr., 1749–1781
32–36.	Johannes Kirch, 1728–†1777
37.	Johann Langenberg, 1642–1657
38, 39.	Probably J. Langenberg
40, 41.	Jacobus Mettman, 1673–†1709
42, 43.	Matthias Medtman, 1635–1665
44.	Philip Wilhelm Marx, 1720–1736
45.	Tönnis (Antonius) Medtman, 1605–1649
46–48, 63	and probably also 65, Bernd (von) Odendal, 1636–1652
49, 50.	Evert Odendal, 1662–1665
51.	Probably Ulrich Odendal, 1653
52.	Wilhelm Odendal, 1640
53.	Unknown
54.	Jacob Römmer (Römer), 1713–1727
55–59.	Petrus Römer, 1738–1750
60.	Huppert Weber, 1662–1664
61.	Michl Grevenberg (18th century)
62.	Unknown
63.	Bernd Odendal (see 46–48)
64.	Jürgen [George] from Metz, 1648–1649
65.	See 46–48
66.	Unknown

Another Cologne master of the seventeenth century has recently been identified, Thoenis Dorfmann, whose sign is a D superimposed on a T.

171

Cologne weight and scale makers whose mastersign has not yet been identified:

Johan Altenberg, 1848–1865
Chr. Hubert Altenberg, 1870
Johann Bamberg, 1780–1781
Engel Bechem (Berchem), 1655
Franciscus Constantinus Brewer, 1718–1728
J. B. Coellen's widow (19th century)
Joh. Everhardt Coellen, 1822
Joannes Fuchs, 1756
Anton Caspar Grevenberg, 1781–1797
Johan Conrad Gussen (apprentice, 1762–1781; master, 1781)
Johan Lützenkirchen, 164?
Heinrich Lützenkirchen, 1655
Johann Baptist Mager (1831–1866)
Matthias Mager, 1777–†1826
Hermann Medtmann, 1650–1674
Paulus Mettman, 1708–1709
Theodor Prümm, ca. 1800
Tonnes Reidt, 1649–1657
Werner von Reidt, 1625
Master Schouff, 1611
Thomas Wagemacher, 1555

Forien de Rochesnard and Lugan (1955) also mention the following dates for weight makers of Cologne, without giving a source. They were therefore not included in the list above:

Bernd Odendal, 1636
Jean Lützenkirchen, 1649
Werner von Reidt (16th century)

Mastersigns of Box Makers for Gold Scales

Key to Chart 4. This chart contains all the signs known to this author of the masters who made boxes for gold scales (*Laden*). Unfortunately, not one of these signs can yet be attributed with surety to a definite person. The only clue to where, when, and by whom the wooden box was made is the inscription of the balance maker who used and sold this particular box. From information of this kind assembled from thirty-six boxes, the following can be stated (Kisch, 1960b).

Chart 4. Mastersigns of box makers for gold scales.

All the boxes with the mastersigns of their makers still intact, and known to this writer, were used between 1590 and 1662. In the sixteenth century they were used exclusively by weight and scale makers in Cologne (1590–1611) and only later, up to 1662, by Dutch artisans also (but not exclusively), mainly from Amsterdam.

Table 10 shows that no boxes were imported from Holland into Cologne, but there was apparently a moderate export from Cologne to Holland. Within a certain degree of probability it can be concluded from Chart 4 that the mastersigns 1, 4, 5, 8, 9, 10, 11, 20, 24, 26, 27, 31, 35 belong to Cologne masters and 7, 13, 16, 17, 18, 22, 23, 29, 30, 32, 33, 34 to Dutch masters (probably Amsterdam). Mastersigns 2 and 3 could be of French origin. Signs 6 and 28 were found on boxes from Nuremberg of the eighteenth century.

Most of the signs in the following charts were originally surrounded by a wreath of leaves or a circle of dots which have been omitted to make the arrangement clearer. A date in the sign on a coin weight is the date of its issue, but in other specimens the same mastersign may be accompanied by a different year. In the key to the figures all known years of issue for this mastersign are mentioned and also the collections where they were found (abbreviations for collections are on p. xix). When a certain edition or date is known only from literature, the source is quoted. The book by Zevenboom and Wittop Koning (Z&W) is a valuable source for Dutch weight and scale makers and their signs, as are the monographs of Dieudonné (Dd) and Forien de Rochesnard and Lugan (F&L) for France, and Sheppard and Musham (S&M) for England. All are so complete and informative that only the mastersigns not mentioned in these sources are included here, unless a great deal can be added to the already known material, or when, as in the catalogue of F&L, the masters are mentioned but their signs not reproduced.

The charts identify the objects on which the signs are found and are particularly valuable for identifying the signs of the coppersmiths and weight makers of Nuremberg. Our knowledge in this field is based on the important paper (1915/18) by Stengel (S), but many specific data have been added from inscriptions on the boxes for gold scales found in collections all over the world and from adjuster's marks with dates stamped on weights. Of course one must realize that the adjuster's date on a nest of weights from Nuremberg, for example, does not necessarily mean it was issued that year by the master whose sign appears on it; it indicates only that the object cannot have been made later than the given date.

Stengel's excellent monograph contains the most comprehensive and scholarly listing of mastersigns of the coppersmiths of Nuremberg and their owners up to this time, and Charts 5 and 6 therefore list only the mastersigns which the present author has found on various objects. This seemed useful

Chart 5. Mastersigns on Nuremberg nested weights.
Continued on next page.

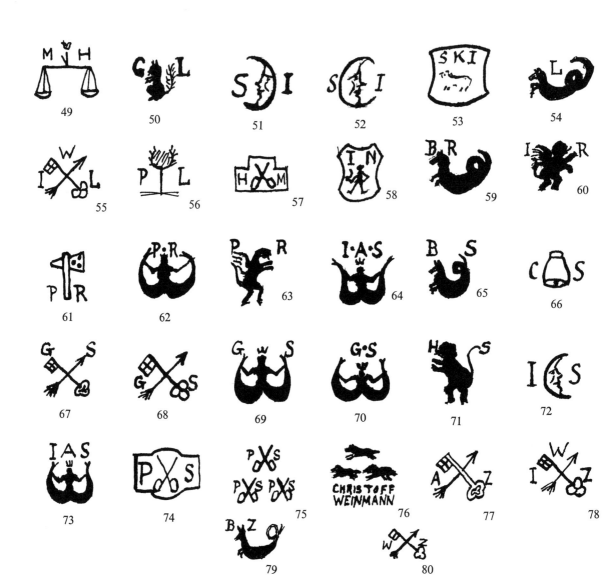

49   50   51   52   53   54

55   56   57   58   59   60

61   62   63   64   65   66

67   68   69   70   71   72

73   74   75   76   77   78

79   80

82   83   84   85   86

87   88   89

because Stengel provided a very limited number of illustrations. It is known that the same mastersign was sometimes used in Nuremberg for generations by different masters, and consequently it is often difficult to date an object, although it can sometimes be ascertained approximately, especially with nested weights, if the adjuster's mark bears a date. In the following listing only the earliest dated mark is reported. The fact that successive masters often used the same mastersign accounts for the fact that the initials of the producing master are not necessarily the same as the initials on the mastersign.

Chart 5 contains only the signs found on nested weights and Chart 6 those on other weights and on scales. The same mastersign may therefore appear occasionally in both charts.

In the following key my great indebtedness to the works of Stengel and of Zevenboom and Wittop Koning is abundantly clear.

Key to Chart 5. Shown here are the mastersigns on Nuremberg nested weights. Numbers 1 to 80 refer to the number of each sign in this chart. The number is followed by my description and by an S followed by Stengel's number for the sign, in parentheses. If the S is followed by 0, the sign is not in Stengel's list. Next come the various users of this mastersign with year(s) of their activity according to Stengel. Then follow the location (see Collections and Abbreviations, p. xix) and, if known, the inventory number of the object bearing this mastersign and the weight of the entire nest of weights; also, as far as possible and in square brackets, the oldest sealer's mark on this set, which may furnish some clue to the maker. References to the tables in Zevenboom and Wittop Koning's book (Z&W) are given at the end of some entries.

℔ = pound; M = mark = $\frac{1}{2}$ ℔; L = lot = $\frac{1}{32}$ ℔; † = year of death

1. Clover	(S-81). (i) Hans Gscheid †1540; (ii) Sebastian von Ach †1571; (iii) Sebald Gscheid, 1567, 1597. CtYSt: CIN-4, 4 ℔; CtYSt:CIN-5, 4 ℔. LSc [1661], 32 M. CoUl, 1 ℔. ZL:22405, 1 ℔. NYBK, 2 sets, 8 ounces each.
2. Grapes and horn	(probably S-208). Friedrich Mend †1630. StAD[1554].
3. (?)	(S-0). CoUl, 8 ounces.
4. Three hearts	(S-62). (i) Christian Engelhart Beck 1695–; (ii) Tobias Martin Kolb; (iii) Matheus Siegler, son-in-law of Kolb, 1787–; (ii) and (iii) as a rule with Kolb's initials. NYBK, 4 ounces.
5. Bow and arrow	(S-19). (i) Georg Bernhard Weinmann 1656–1685; (ii) Leonhard Weinmann, son of (i), 1693–1716. ScL, 4 M. BrAC. Z&W-19.

177

6. Horseshoe   (S-72). No weight makers. NGN:WI 399.

7. Spigot and clover   (S-0). EMA, 2 ℔ [1680].

8. Three spigots   (S-0). MzA, 8 ℔ [1770].

9. Lock   (S-169). (i) Hans Wilhelm Weinmann 1656–; (ii) his son Hans Jochum Weinmann 1680–; (iii) Erasmus Fleischmann 1711–; (he bought the sign from the widow of Weinmann, the son). Z&W-23. GAH, on nest of unusual form as later made in Naples, 4 ℔. LWH:R18648, 6 ℔. WiAU. CtYSt:CIN-9, 4 ℔. NYCD [1787]. CR:110, 4 ℔ [1774]. IT, 32 ducats. KnF, 4 ℔. BtMu, apothecary weights.

10. Three locks   (S-0). GAH:1258, 8 ℔.

11–14. Goblet   (S-80) (i) Georg Fleischmann 1667–; (ii) Johann Erasmus Fleischmann 1727–; (iii) Johann Reinhart Lenz 1766–1795 (received the sign from his brother-in-law); (iv) Christoph Lenz 1796–. Z&W-3a, 3b. Many nests are extant with one of these four signs. Probably the four typically different kinds were used by different masters.

11.   CtYSt:CIN-16, 1 ℔ [1785]; CtYSt-CIN-17, 12 ℔. PCN:17255, 16 ℔; PCN:8080, 8 ℔. NYBK, 1 ℔.

12.   UUF:F286, 2 ℔ [1741], PCN:4256, 2 ℔; PCN:17572, demimark. LScC possesses 15 nests with sign 11 or 12, most of them 1 ℔, three only 1 M, from beginning of nineteenth century. On one only is impressed the year [1617], which may be a misprint for 1817. NYCD:96, 16 ℔ [1807]; NYCD:93, 4 ℔.

13.   LWH:4438, 12 ounces, also R18664, 16 ounces; R18666, 8 ounces [1830]; R118668, 16 ounces [1840]. NYBK, two each 16 ounces and one 4 ounces. LScC, 16 ounces. ZL:LM 22404, 36 ℔. LM 11201 and 11202, each 16 ℔; LM 22406, 1 ℔.

14.   CtYSt:CIN-15, 16 ℔.

15. Three goblets   (S-0). CtYSt:CIN-18, 32 ℔.

15a. Crucifix   (S-88). (i) Georg von Ach, 1656–; (ii) Georg Jacob von Ach; (iii) Friedrich Holzmann (cousin of Georg J. von Ach) got it from (i) 1697–; (iv) Joh. Georg von Ach –1790; (v) Meister Fleischmann 1800–. (BxRA has a nested weight, 1 ℔, with this sign.)

16. Two crossed spears   (S-180). (Wolfgang) Singer 1800–. Z&W-24. LScC, 1 ℔. BxRA, 16 L.

16a. Two crossed syringes	(S-182). (i) Victor Abend 1791–; (ii) Johann Jacob Pabst 1799–1805; (iii) Georg Pabst 1814–. BxRA, 16 L.
17. Battleax	(no initials) (S-0). Z&W (see our No. 61). LWH:64, ℔ or mark [1767]. CtYSt:CIN-6, 16 L [1754].
18, 19, Scale	(S-204). Johann Caspar Wild 1795–1803 or 1804. Z&W-1. No. 18 and No. 19 are so different that they probably were signs of different masters. The one listed as (S-204) was definitely not the only one using this symbol. Z&W mention two nests, one marked [1726], the other [1765].
18.	LeRN, 16 ℔ [1764]. NYBK, 8 L. BxRA, 1 ℔. KnF, 8 ℔.
19.	CtYSt:CIN-8, 1 ℔. CIN-7, 12 ounces. NYBK, 16 L.
20. Open crown.	(S-91). David Hoppert 1791–. The sign was also used earlier. NYBK has two nests with the open crown from the 16th-century type; one bought in Florence has in each cup the Medicean hallmark with the year 1578. NYCD, probably also 16th century. ZL, 16 ℔ [1699]; also Dep-2289, 32 ℔.
21. Three open crowns	[not to be mistaken for Swedish adjuster marks also having three crowns (S-92)]. (i) Joh. Georg Loos 1758– (he got it from his deceased brother-in-law); (ii) Carl Gottlieb Lorenz 1795–. CtYSt has three such nests, one has 16 and two each 8 ℔. KnZ:D-80, 16 ℔ [1602] (last sealer's mark 1774). StAD:IT [1769].
22. Two keys crossed	(S-171). (i) Johann Conrad Schön 1781–; (ii) his son Christoph Martin Schön 1794–. HdA, 0.382, 16 ℔ [1791].
23–26. Arrow and key crossed	(no initials) (S-172). (i) Conrad Weinmann 1604–; (ii) Georg Schüller (Schiller) 1656–; (iii) Andreas Ziegengeist 1681– (he used it after his cousin Georg Mittmann in 1681); (iv) Johan Wolf Zickengeist 1721– (he got it from his stepbrother Andreas Zickengeist with initials; see No. 78). The typical differences between 23, 24, 25, and 26 apparently indicate different masters. Z&W-21.
23.	KnGS. BxCMe, 1 M. BxRA, 1 ℔.
24.	LSc. WiAu [1724]. GAH:8166. ZL:7611a, 32 ℔, 4 ℔ [1699].

25.	PCN:819, 4 ℔.
26.	CtYSt-CIN:26, 16 ℔. BxRA, 4 ℔.
27. Head of Negro	(S-126). (i) Georg Weinmann †1604; (ii) Hans (Christoph) Zickengeist 1674– (he bought the sign from Mrs. Seufferheld); (iii) Hieronymus Ziegengeist; (iv) Georg Ziegengeist 1720–(with permission of his brother Hieronymus Ziegengeist); (v) Leonhard Hauerstein 1781–. Z&W-17. LWH:A 29678. HlM:A 8168. ZL:7011. NYBK, 17th-century type, 8 ounces. CtYSt:CIN:1, 4 ℔. BxRA, 1 ℔.
28. Three heads	(apparently not Negroes) (S-0). MAN:1492, 5942.
29. Man with tree	(S-108). (i) Georg Lorenz Braun 1674–; (ii) his son Johann Paulus Braun 1719– (he used his father's sign with the initials J.P.B.). Z&W-25 know this sign of the son with year [1703], UUF:F 285, 4 ℔ [1725], and 1 ℔ [1707]. ZL:LM 7011a, 4 ℔. BxRA, 1 ℔.
30. Three men with trees.	(S-0). CoUl, 16 ℔.
31, 32. Mermaid	(no initials) (S-123). These two are so typically different that they were probably used by different masters. Z&W-27. (i) Jonas Paulus Schirmer; (ii) Hans Andreas Schmid 1699 or 1700–; (iii) Christoph Schön (Schem) 1727–1730 (he got the sign from his teacher Schirmer); (iv) Georg Scherb 1730–; (v) Paulus Ritter; (vi) Paulus Frühinsfelt 1768– (he got the sign from Paulus Ritter); (vii) Martin Christian Schön 1787–1794 (he bought the sign from the widow of Frühinsfelt); (viii) Johann Jacob Spagel 1796–. Z&W-27.
31.	NGN:WJ 187 [1770]. BrAC, 2 ℔. BxRA, 2 ℔.
32.	NGN:WJ 424, 16 ℔ [1809]. PCN:17198, $\frac{1}{2}$ M. CtYSt:CIN-20, $\frac{1}{2}$ ℔. NYCD:94, 8 L. ZL:LM 24778 [1699]. PaH. BxRA, 8 L.
33. Rampant horse	(S-157). KnZ:D 85.
34. Standing bear	(S-9). (i) Stephan Weinmann; (ii) Hans Jacob Trautner (inherited this sign); (iii) Georg Leonhard Weinmann 1728–1730 (inherited sign from his father); (iv) Joh. Jac. Wilt 1766 or 1767 (bought the sign from Lucas Weinmann and his mother). Z&W-2. NYBK, 8 ounces, 1 ℔. BxCMe, 8 M, 16 M.

180

35. Seahorse (S-124). (i) Christoph Schön 1746–; (ii) Joh. Conrad Schön 1750–; (iii) (Gottlieb) Heinrich Wild 1794–; (iv) Joh. Caspar Wild 1822–. Z&W-29. MG [1804]. CtYSt:CIN-13, 8 ounces. CIN-14, 12 M [1788]. NYBK, 2 nests each 1 ℔. LScC, 24 ounces. LSc:1937–170. CoUl, 2 M. BxRA, 1 ℔, 16 L.

36. Griffin rampant to the right (no initials) (S-50). (i) Conrad Most; (ii) (Joh.) Sebastian Küntzel after 1707; (iii) Paulus Ritter; (iv) Christof Wiliwalt Schick 1766–1769 (he got the sign from his cousin P. Ritter); (v) Master Fleischmann 1800–. Z&W-7, 7a, 7b. BxCMe, 16 L. NYCD, small nest. NYBK, 8 ounces.

37. Cock (S-52). (i) Christoph Hainlein 1670–; (ii) Christoph Jobst Stohdruberger 1788– (he got this sign by marriage); (iii) Johann Zacharias Abend 1820–. Z&W-8. The earliest nest with this sign is dated [1776]. BxCMe, 8 L. BxRA, 2 specimens, 1 ℔, one 2 ℔. NGN:WJ 423, 16 ℔ [1774]; NGN:WJ 1644, 8 ℔ [1796]; NGN also possesses two apothecary scale boxes, each with a nest with this sign, one marked [1766], the other [1779]. MG has three nests with this sign marked [1804]. KnZ:D 102. LWH:A 172578. LSc:LC, one nest [1818]. CoUl, 16 L. LSc, 1 M. NYBK, 8 ounces.

38. Stork (no initials) (S-189). (i) Lenhard Abend 1667–; (ii) Hans Lönhart Abend 1707– (he got the sign from his father); (iii) Georg Abend 1765– (got it from his father); (iv) Johann Augustin Abend 1804–; (v) Georg Abend 1826–. Z&W-18. IT, 32-ducat nest. BxRA, 8 L.

39. Lamb with flag (S-93). (S mentions three nests from the 16th century.) (i) Sebastian Küntzel 1667–; (ii) Joh. Sebastian Kintzel 1707–; (iii) Jacob Wilhelm Reider 1722– Z&W-14. CoUl, 32 ducats. KnZ:205, 2 ℔ [1684]. IT, 8 M.

40. Deer (S-67). (i) Hans Wolf Herolt 1649? 1667–; (ii) Erasmus Herolt 1693– (he bought the sign and the house from his cousin Hans Wolf Herolt). PCN:3262, 64 M. CtYSt:CIN-10, 16 ℔.

41. Wolf (S-211). Christoff Weinmann 1667–. Z&W-26. CtYSt:CIN:11, 32 ducats. ZL:18349, 4 ℔ (as in Z&W-26 the initials C over W in a circle are added).

42. Stork
    G A

(see No. 38; S-189, Z&W-18). Georg Abend 1765–. NGN:WJ 710 [1800], 64 M. NYBK, 16 L. GAH: 11371, 16 L. BxRA, 1 ℔, 4 ℔.

43. Stork
    L A

(see No. 38; S-189, Z&W-18). Hans Lönhart Abend 1707–. KnZ, 4 ℔ [1734]. NYBK, 1 ℔ [1789]. ZMI. LWH:R 14751, A 85593, 8 M. GAH:13424. CtYSt: CIN-24, 4 ℔. BxRA, three specimens (1, 1, 2 ℔.) also 4 ℔.

44. Stork
    Z A

(S-0). NGN:WJ 701

45. Arrow and key
    crossed
    G S

(S-172). See No. 25 but initials GS (or GB) added. Georg Schüller 1656–. Z&W-21. PCN:17196, 8 ℔ [1663].

46. Seahorse
    H B

(see No. 35; S-0, Z&W-0). NYBK, 1 ℔.

47. Clover
    I F

(S-0). BsPH, 8 ℔.

48. Negro head
    H Z G

(see No. 27; S-0). NGN:WJ 974, probably by Hans Ziekengeist 1674. SN, 32 M.

49. Balance
    M H

(see No. 18; S-0). KnZ:D 103, 8 ℔. BsPH, 8 ℔, also 16 ℔ (on this nest sign No. 49 is impressed three times). BxRA, 8 ℔.

50. Squirrel
    C L

(S-26). (i) Caspar Leitner 1689–; (ii) Martin Harburger 1709–; (iii) Joh. Paulus Dimmler 1730– (he bought the sign from Harburger). EMA, 2 ℔.

51. Half-moon facing
    left
    S I

(S-127). (i) Veit Hoffmann 1569–; (ii) Enderas Blechner †1570; (iii) Lonhard Hoffmann 1656–; (iv) Erasmus Fleischmann 1711–; (v) Georg Martin Fleischmann 1714– (he got the sign from his brother Erasmus); (vi) Stephan Jäckel 1769–; (vii) Andreas Philipp Frühinsfeldt 1810–. Z&W-15. ZMI.

52. Half-moon facing
    right
    S I

(see No. 51; S-127, Z&W-15). NGN:WI 700, 1 ℔ [1798]. LWH, 2 ℔. CtYSt:CIN-19, 2 ℔. LScC, 1 ℔. CoUl, 2 sets, each 16 ℔.

53. S K I over
    animal(?)

(S-0). CtYSt:CIN-12.

54. Seahorse
    L

(see No. 35, 46). UUF:F280, 1 M.

55. Arrow and key
    crossed
    I W L

(see No. 24; Z&W-21). LWH:A 181786, 4 ℔ (the third letter could be a Z in which case the nest may be like Z&W-21 from Johann Wolf Zickengeist 1721).

56. Tree P L	(S-0, Z&W-4). LWH:62959, 8 ℔.
57. Scissors H M	(S-0, Z&W-0). KnZ:D 131, 8 ℔.
58. Man with sickle I M	(S-119). Joh. Friedrich Meyer 1781. IT, 32 ducats.
59. Seahorse B R	(see No. 35, 46, 54; S-0). PCN:17571, 16 M. BxRA, 2 ℔.
60. Griffin I R	(see No. 36). GAH:261, 4 ℔ (the first letter could be P, in which case the set could have been made by Paulus Ritter 1756).
61. Ax P R	(see No. 17; S-0, S&W-9). NGN:WJ 1642, 64 M. LWH:R 18653, 4 ℔. BxRA, 4 ℔.
62. Mermaid P R	(see No. 32; S-123). CtYSt:CIN-21, 8 ℔. ZL:11199. HeHM, 16 ℔. Probably Paulus Ritter, 18th century.
63. Griffin P R	(see No. 36; S-50, Z&W-7). Probably Paulus Ritter, 18th century. UUF:F 291, 4 ℔. KnZ:D 124, 8 ℔ [1756]. LWH:R 18661, 4 ℔. BxRA, two specimen, each 4 ℔.
64. Mermaid I A S	(see No. 32; S-123, Z&W-27). CtYSt:CIN-23, 64 M [1729]. Probably Hans (Johann) Andreas Schmid 1700(?).
65. Seahorse B S	(see No. 35, 46). MzA, 16 ℔ [1751].
66. Bell C S	(S-48). (i) Christoph Schirmer 1656– after 1690; (ii) his son Jonas Paulus Schirmer 1715; (iii) Georg Scherb. Z&W-10. LSc, impressed on set three times [1661] (probably by Christoph Schirmer). KnF, 1 ℔.
67. Arrow and key crossed G S	(see No. 24, 55). Probably Georg Schüller 1656–. MAN, 16 M. CtYSt:CIN-25; CIN-26, 16 ℔; CIN-28, 8 M. NYBK. BxRA, three sets, 4℔, 1 ℔, 1 ℔.
68. Arrow and key crossed G S	(see No. 67). LSc, 16 ℔ (with this sign marked three times on the top).
69, 70. Mermaid G S	(see No. 32, S-123, Z&W-28). Probably Georg Scherb 1730–. WB:G 4188. KnZ:D 144-16. PCN: 1719-7, 4 ℔. CtYSt:CIN-22, 4 ℔. SN, 32 M [1742]. BxCMe, 1 ℔ BxRA, 1 ℔.
71. Rampant lion H S	(S-102, but no initials). (i) Hans Georg Fink 1691–; (ii) Christof Löblein 1735. PaH:A534, 8 M.

72. Half-moon facing right
I S
(see No. 51, 52; S-127, Z&W-15). MG [1804].

73. Mermaid
I A S
(see No. 64). Probably same Master Hans Andreas Schmid. BsPH, two sets, one 4 ℔, one 8 ℔.

74, 75. Scissors
P S
(see No. 57; S-164). (i) Jonas Paulus Schirmer; (ii) Georg Leonhart Weinmann 1730–; (iii) Joh. Melchior Gehr 1801– (he bought the sign from Weinmann). Z&W-20.

74.
NYCD:95, 64 M.

75.
FmK, 32 M.

76. Three wolves
CHRISTOFF
WEINMANN
(S-211). Christoff Weinmann 1667–. KnZ:R M 20629 [1672].

77. Arrow and key crossed
A Z
(see No. 67, 68; S-172, Z&W-21). Probably Andreas Ziegengeist 1681–. PCN:18557, 8 ℔. GAH:8165. HoW, 16 ℔ (this set is from Batavia; it bears a note in old writing saying *Andreas Ziegen . . .*) BxRA, 8 ℔.

78. Arrow and key crossed
I W Z
(see No. 67, 68, 77; S-172, Z&W-21). Johann Wolf Ziegengeist 1721–. UUF:F 292, 4 ℔. MAN:7097, 32 ℔.

79. Seahorse
B Z
(see No. 35, 46, 65; S-0). BxRA, 1 ℔.

80. Arrow and key
W Z
(see No. 67, 68, 77, 78; S-172, Z&W-21). BxRA, 4 ℔ (probably Wolf Zickengeist 1721–).

81. Angel with sword on dragon
(S-0). PCN:3263, 32 M [1710].

82. Griffin rampant to the left
(S-0). See also No. 36 which probably is different from the sign of the master in S-50 (griffin rampant to right side). Both extant nested weights with this sign are very large. SN has a beautiful one of 32 ℔ bearing the adjusters' marks of [1664], [1666], [1729], [1815]. BxRA has a similar magnificent one of 32 ℔.

83. Eye
G K
(S-8). (i) Georg Künstel (Küntzel) 1688–1767; (ii) Joachim Künzel 1735–; (iii) Johann Phil. Richter 1794–; (iv) Joh. Rosenschon 1816–. BxCM has a nested weight of 4 ℔ on which this sign is twice impressed, also the year [1742].

84. Man with tree
I B B
(see No. 29, 30; S-0). BxRA, 1 ℔.

85. Mermaid
P K

(see No. 32, 62, 69, 70; S-0). BxRA, 4 ℔. One 17th-century type (4 ℔) with an old Spanish sealer's mark was seen at a New York dealer's shop.

86. Lamb with flag
S K

(S-93, but without letters). (i) Sebastian Küntzel 1667–; (ii) Joh. Sebastian Kintzel 1707–; (iii) Jacob Wilhelm Reider (Reuter) 1722–; (iv) Joh. Leonhard Niedel (Riedel) 1728– (he bought this sign from Reuter); (v) Conrad Rossner before 1766; (vi) Joh. Albrecht Schon 1766– (he got the sign from Conrad Rossner). BxRA, 1 ℔ (probably by Sebastian Küntzel).

87. Scale

(see No. 18, 19; S-204). Similar to No. 18 but definitely different balance form. BxRA, 1 ℔.

88. Crossed key and arrow
G M

(see No. 22–26; S-172). SN, 16 ℔, adjusters' marks [1685] and [1825]. Probably the sign of Georg Mittmann whose widow gave it in 1681 to her cousin Andreas Ziegengeist.

89. Wolf
C W

(see No. 76; S-211). (i) Christoff Weinmann 1667–; (ii) Johann Christof Zeltner–1711. BxRA, 2 ℔.

Key to Chart 6. The following abbreviations for the scales and weights (other than nested weights) on which mastersigns or adjusters' marks appear are used in the key to the chart:

○   One round pan of gold scale marked
▽   Triangular pan of scale marked
○○   Both round pans marked
Ⓝ   The round pan marked with arms of Nuremberg (it is usual for Nuremberg balances to have the Nuremberg coat of arms in one pan and the master-sign in the other, which furnishes provenance).

One pan of the so-called ducat scales (see p. 47) usually bears the symbol of the Hungarian ducat, once a very popular gold coin; the other bears the mastersign of the maker. Here, too, it is safe to state that the master worked in Nuremberg where nearly all the extant ducat scales originated. A question mark is appended to date and places that are assumed but cannot be documented.

Chart 6. Mastersigns on Nuremberg scales and weights other than nested weights.

1. Half man with hammer
G W A

(S-0). NYCD:105, in box by Joh. Friedr. Mayer on ▽ of scale. IT:1234, in ○ of scale.

2. Crosier
I A

(S-17). (i) Hans Christoph von Ach 1649–; (ii) Hans von Ach 1719–; (iii) Joh. Jacob Bohm 1765–, who got the sign from (ii). LWH:95, in lid of box [1675]. In ○; in the other Ⓝ. Probably Hans (Johan) von Ach.

3. Crosier
I V A
1742

(see No. 2; S-17). KDN:GP 1810, ducat scale. Sign impressed on lid, Ⓝ. Hans (Johan) von Ach.

4. Crosier
N C V A

(see No. 2; S-17). SN:21860, Nuremberg-type weight box. Sign impressed on box. Probably Hans Christoph von Ach.

5. Man with sickle
I I B

(S-0). BsH:1914/171, scale, in one ○ this sign; in the other Ⓝ.

6. Heart
W C
H

(S-0). NH:958, on lid of box also [1761]. An L in ▽ and ○.

7, 8. Wheel
A D

(S-152). (i) Balthasar Wagner 1667–; (ii)——Deinert; (iii) his son Joh. Conrad Deinert 1778–.

7.

NGN:WJ 719. On one ○ of ducat scale is AD. WB:G 3713, in ○○. NYNS, in both ○○. IT, in ○○. WB:G 3711, in ○ of ducat scale.

8.

WB:G 4185, impressed on lid of small weight box.

9. Two facing doves

(S-194). NGN, on label of ducat scale, also *Friedrich Däuber*. This same label with the name is in two ducat scale boxes in SN. Also IT:464. NH. PaH: 10106.

10. Dry cup
L D

(S-101, but no initials). (i) Jacko Kinstel [Küntzel] 1667–; (ii) Johann Loss 1732–; (iii) Paulus Deinert 1758– (he got the sign from his brother-in-law); (iv) Joh. Gabriel Bach 1809–. BaW, on brass lid of Nuremberg box.

11. Dry cup
P D

(see No. 10). (Paulus Deinert). NGN:WJ-1124, WJ = 1930. BsH. LWH. WM-2. LeRN:A 400. KnFZ. CtYSt:B-15, 115, 133. NH, NYBK. BrH. All on brass lid or lid of boxes.

12. Dove
P D

(S-0). ZMI, in ○○ of scale; box signed with No. 28.

13. Horseshoe
W D 1591

(S-0). LSc:1935–560, on ▽.

187

14–16. Barrel
    M F

(S-36). Max W. Fleischmann 1717–. No. 14, ZMI, on ▽. No. 15, LWH:A-61543, on ▽. No. 16, LWH:A-70598, in ▽ of Cologne weight box of 1651.

17. Three hearts
    G H
      B

(S-62). (i) Christoph Engelhardt Beck 1695– (he was probably a balance maker like two of his ancestors); (ii) Tobias Martin Kolb; (iii) Matheus Siegler 1787– (he got his sign from his mother-in-law). ZI, on a 1-ounce weight.

18–20. Scepter
    I G H
(and also scepter
without initials)

(S-0). No. 18, NYNS, on ▽ in Dutch box of 1696. SN:40997, in ▽ of small Nuremberg box, 18th century. No. 19, NGN:WJ 395, on ducat scale in ○. No. 20, WB:G 3728 on ducat scale in ○. NH, on two ducat scales in ○.

21. Crossed swords
    G I

(S-0). LWH; in a box of Paulus Deinert this sign is on each weight.

22–26. Three hearts
with initials or
without

(see No. 17; S-62). No. 22, WB:G 3715, in box: *Tob. Martin Kolb* and 1770, Ⓝ. No. 22 in ▽. CtYSt: B-62, in box label: *Tobias Martin Kolb* and 17... No. 22 in ▽, Ⓝ. No. 23 (S-0), WiAu, in ▽, Ⓝ. No. 24, ZMI, sign on a 1-ounce apothecary weight. No. 25, NGN:WI 708, in box of 1767, sign on ▽, Ⓝ. SN, label in box: *Tobias Martin Kolb 1773*, sign in ▽, Ⓝ. CtYSt:B-116, on brass lid: *Kolb 1767*, sign on ▽, Ⓝ. NGN. No. 26, NGN:HM, in ▽, Ⓝ; also NH in ▽. CtYSt:B-13, label: *Tobias Martin Kolb 17* . . .; also sign 26 but *T.M.K.* and the hearts arranged as in No. 38.

27. L

(S-0). NGH:HM ducat scale, label: *Christoph Lowlein*; NH:958 in ▽ and ○ [1761].

28–30. Dry cup
    I L

(see No. 10, 11). IT:1237, ducat scale, label of Johann Los. No. 30, same sign on ducat scale at PaH and label of Paulus Deiner in box. No. 28, CtYSt:B-10, impressed on box, also *1740*. LWH: 40540. NGN:WI 403 [1553]. ZMI. NYBK. PaH, on box, also *1774*. No. 29, NH, on label of ducat scale, also [*Johan*]*n Los*. Also NGN. HeHM.

31. In circle C M
            8

NGN:WI 304, on ○ of ducat scale. LSc:1946–109. NH:828, in ○○. NYBK, in ○○.

32. Scepter
    C M

(S-0). SMk:SHM 23532, in ○ of ducat scale.

33, 34. Man with sickle
I F M

(S-119). (Joh.) Friedrich Meyer 1781–. NGN:WJ 1788, SWB, in ▽ and Ⓝ. GN has three other boxes with mark 33 on ▽, in one the year 1764, in the other 1782. No. 34 is in ○○ of box at NYNS, on the label: *Joh. Friedr. Mayer*, the *verpflichteter Waag und Gewicht Eichmeister in Nürnberg 1805*. NYBK, three weight and scale boxes with sign on ▽ and Ⓝ. NYCD, ○ and Ⓝ.

35. Star
T R

(S-0). KDN, in ▽ [1653].

36. Man with hoe
E S

(S-109). (i) Master ES; (ii) Friedrich Preuss 1705–; (iii) Johann Sigmund Michel 1792–. WB:G3710, weight and scale box, in lid: *Nürnberg Anno 1729*, on ○○.

37. Caliper
G S

(S-213). (i) Johann Gabriel Sigler 1773 or 1774 (got the sign from his father); (ii) — Siegler 1801–; (iii) Leonhard Dietz 1815–. LWH: 140843 on the little brass door in typical Nuremberg weight and scale box, 18th century.

38. Three hearts
G H S

(see No. 22, 23; S-62). ZL on a 1-ounce pharmaceutical weight.

39, 40. Two crossed arrows
I S

(S-145). (i) Hans Neydel †1549; (ii) Philipp Winter 1648– (he got permission from Stefan Forster to use this sign); (iii) Master HL; (iv) Johann Sichler 1739–. No. 39: NGN:HM on ○ of ducat scale. KDN:GP-1926, on ▽. BsPH:P-7.138, on ▽. LWH, on ○ in ducat scale. No. 40; SN, on ▽; *1620* engraved on lid of this box. No. 39 and 40 are apparently the signs of Johann Sichler.

41. Three crossed arrows (no initials)

(S-146). Georg Seidel 1771. LSc:1937–170, large pharmaceutical weight and scale box, in ○○ sign No. 42. In the box sign No. 41 and the name *Seidel* are impressed.

42. Three crossed arrows
G S

(see No. 41; S-145). SN, on ○.

43. Dry cup
I G S

(see No. 10, 11, 28–30; S-101). NYNS, on grain-weight door of typical Nuremberg weight and scale box. NYBK, two small weight and scale boxes. No. 43 on the lid of each. CoUl, a similar box. BsH, pharmaceutical weight and scale box with sign No. 43 on brass door of grain-weight recess.

189

44. Scepter
    C W
(S-163). (i) Hans Christoph Weber 1694–; (ii) Georg Jacob Himmler 1730– (he bought the sign from "Frau Weber"); (iii) Georg Abend; (iv) Johann Jacob Wild 1781–; (v) Johann Augustin Abend 1795–; (vi) (Christoph) Geissler 1804–. SN, on lid of small weight scale box of the Nuremberg type (apparently) *Hans Chr. Weber.*

44a. H W
(S-0). BrH, in ▽ of old gold scale.

45. Scepter
    I C W
    1704
(see No. 44. S-163). This sign is apparently that of Hans (Johann) C. Weber. NGN:WJ 404 impressed on lid of weight-scale box. IT No. 1233 in ▽ and also IT:1235 in ◯◯.

46. Two crossed
    arrows
    P W
(see No. 39, 40; S-145). Apparently Philipp Winter 1648–. SN, on ▽. NYBK, on ◯ and ▽.

47. Lock
(see No. 9 of Chart 5; S-169). BtMu, pharmaceutical weights, one of ℔s ($\frac{1}{2}$ ℔), two specimens of ʒ, and three of $\frac{1}{2}$ʒ, which bear this sign and also the hallmark of Nuremberg.

48. Chalice
(see No. 11–14 of Chart 5, S-80). NYBK, brass weight with handle, 4 ℔. ZL:LM-14542, sign 48 on two pharmaceutical weights of $\frac{1}{2}$ ounce each.

49. C K
    W
CtYSt:B-15; these letters in a shield are in one ▽ of a balance and also the year 1647; Ⓝ. The sign is also on a coin weight in NYBK.

ANTWERP MASTERSIGNS

Key to Chart 7. The data given by Forien de Rochesnard and Lugan (F&L) are in parentheses; 0 means unknown to F&L; – means that only the initials are known to F&L. If F&L is followed by a date(s) it is the year or the first and last years known to F&L for this sign on a weight; numbers in parentheses are years of issue, the century sometimes omitted; *cw* = coin weight; *Nnw* = Nuremberg nested weights; box or *wsb* = weight and scale box.

1.
(F&L-0). LSc, *cw*. NYBK, *Nnw.*

1a.
(F&L-0). Probably old sign for Antwerp. See No. 47, BxCMe.

2.
(F&L-0). NYNS, *cw.*

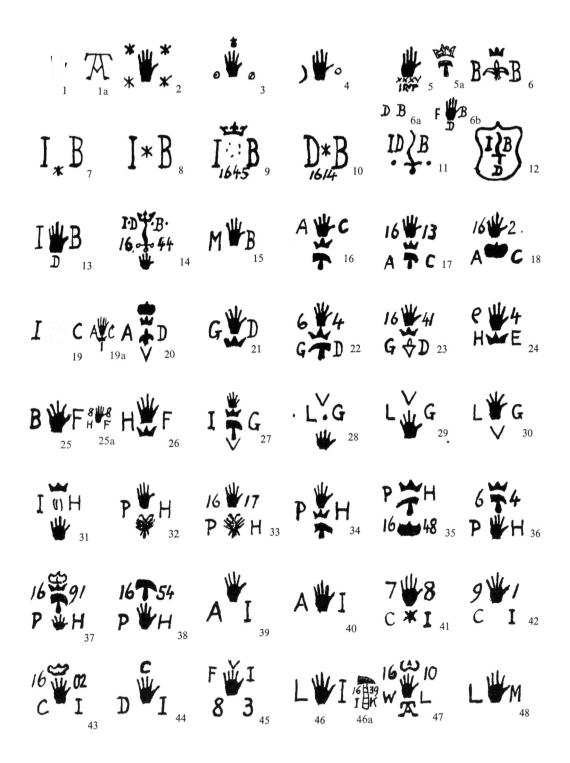

Chart 7. Mastersigns of Antwerp weight makers.

Continued on next page.

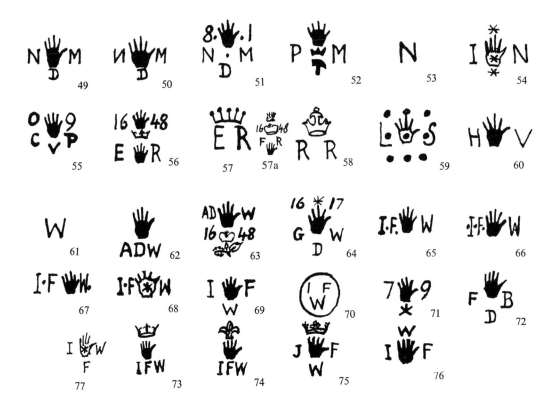

3.		(F&L-0). NYBK, 2 *cw*.
4.		(F&L-0). NYBK, *cw*.
5.		(F&L-0). AbR, *cw*.
5a.		(F&L-0). Aertus van Dunwalt. BxCMe in ▽ of signed box (see p. 195).
6.	BB	(F&L–). BxCMe, *wsb* by Bernard Jaspers Baunnaerts, *opdaude borse inden salm tot Antwerpen, 1649.*
6a, 6b.	DB FDB	(F&L 1766). François de Batist. BxCMe, in box by Fr. de Batist, 1766, *gesworen ijker von Iere Majestÿt.*
7–9.	IB	(F&L-0).
10.	DB	(F&L–). Fr. de Batist. AO:692, pharmaceutical weights, label of Fr. de Batist. BxCMe, box of 1766, with D on ○, B and No. 10 on ▽. LSc:1944:23, on 2 *cw*. AO:689.
11–14.	IDB	Jacques de Backer. No. 11, LSc, on *cw*. No. 12, AO:690. BgG, 1 *cw*. No. 13, PBN:Y21587, in box of J. J. de Baecker 1648. No. 14, PBN:Y2158, *cw* of 1644, 1648, 1669. PCl:1648, 18 *cw* (F&L 1644–1648).

NYBK, 3 *cw*. BxCMe, *cw* of 1644, also *wsb* with stem of Amsterdam and sign No. 8 of Chart 4, in it written: *Gemaecht door Jan de Baeker tot Antwerpen*. On most weights is sign 14 with the year 1669.

15.	MB	(F&L-0). NYBK, on *Nnw*.
16.	AC	(F&L 1610–1619). André Caers. LSc:1944–35, in box of 1630, 3 *cw*. PBN:21587, in box of 1648 on 1 *cw*. PCl:4567, *cw* in box of 1648.
17, 18.	AC	(F&L 1610–1619). André Caers. CtYSt, 3 *cw*. NYBK, *cw*'s from 1613, 1615, 1626.
19.	IC	(F&L 1804). Joseph Theodor Carolus. BxCMe has a *wsb* of 1804 by J. Carolus, *Balantmaecker op D'eymarkt . . . tot Antwerpen*, all weights in it with sign No. 19.
19a.	AIC	(F&L-0). BxCMe, *cw*.
20.	AVD	(F&L only ADW 1648–1650). A. van Dunwalt, 17th century. NYNS, *cw*.
21.	GD	Gerardt van Dunwalt. NYBK, 2 *cw*.
22.	GD	(F&L 1641–1644). Gerardt van Dunwalt. NYBK, 1 *cw* ([16]64). LSc:1944-35, 2 *cw* (1644).
23.	GD	Seen at Paris dealer's. Probably also G. van Dunwalt.
24.	HE(?)	(F&L-0). LSc:1944-23, on 2 *cw* in box of 1677 (4). PBN:FL-380, 1 *cw* (80).
25.	BF	(F&L–). NYNS, 2 *cw*. CtYSt, 4 *cw*.
25a.	HF	(F&L–). BxCMe, 1 *cw* [1588].
26.	HF	(F&L–). NYNS, 2 *cw*, with year (–4, 90). NGN and LScC, *cw* (80).
27.	IVG	(F&L–). PCl, most of the *cw* in box of Gerrit Geens, 1648, also in box of J. de Backer, Antwerp, 1648. NYBK, 2 *cw*.
28–30.	LVG	(F&L 1521). Leonard van den Gheere. CtYSt, *cw*, LSc, 2 *cw*. NGH, 1 *cw*.
31.	IH	(F&L–). PBN:Y21857, 2 *cw* in box of J. J. de Baecker 1648.
32, 33. (double eagle)	PH	(F&L 1617–1645). Pierre Herck I. CtYSt, *cw* (1617). Sc, in *wsb* of 1630, *cw*. PBN, 2 *cw*, LSc, 1 *cw*. One seen at Paris dealer's.
34–38.	PH	Pierre Herck II. No. 34, PCl, *cw* in box of 1648. NYNS, 4 *cw* (1645, 1648, 1649, 1691). A0, on *cw* in box of Peeter Herck. CtYSt, 12 *cw* (1645, 1648, 1649, 1650, 1654, 1655). PB:Y2158, *cw* [16]64.

193

		NYBK, 7 *cw* (1645, 1647, 1648, 1650, 1667). (F&L 1645–1717). PCN, 1 *cw*. BxCMe, *cw* like 37 (1645, 1647, 1648, 1649, 1650, 1667).
39–40.	AI	(F&L–). LSc:1935–560, 2 *cw*. CtYSt, 1 *cw*. PCl, *cw*. NYNS, *cw*.
41–43.	CI	(F&L 1578–1604). Corneille Janssens. CtYSt, 3 *cw*, NYBK, 3 *cw*. PBN:F1375, *cw*. BxCMe, 1 *cw*. At dealer's in Paris 2 *cw* seen. The years on these *cw* were 78, 79, 81, 86, 91, 92, 96, 1600, 1602, 1604, 1606.
44.	DCI	(F&L–0). LWH, *cw*.
45.	FVI	(F&L–0). NYNS, *cw*, 83 in box of G. de Neve 1628.
46.	LI	(F&L–0). CtYSt, *cw*.
46a.	IK	(F&L–0). BgG, Dutch box inscribed: *Gemaect by Jan Caen inde halsteeg int goutgewicht. 1646. cw*, Antwerp.
47.	WAL	(F&L only WL 1596–1619). Wilhelm von Langenberg(?) CtYSt, *cw*. BxCMe, box with year 1598 on label written *Wilhelm von Langenberg* in ○; No. 1 in ▽○. No. 1a on seven *cw*; No. 47 without crown and the year 96.
48.	LM	(F&L–). NYNS, *cw*.
49–51.	NDM	(F&L–). CtYSt, *cw*. NGN, 4 *cw* (81).
52.	PM	(F&L–0). LSc, 2 *cw*.
53–54.	IN	(F&L 1749). Jacobus Franciscus Neusts. AO:687, in box with Neusts' label of 1749. No. 53 on *cw*. PBN, 1 *cw*. NYBK, 1 *cw*. BxCMe, on scale of the box marked *1749*. No. 54, BxCMe, 1749 on *cw*. A0:271, 685, 246, 671. CtYSt, *cw*.
55.	CVP	(F&L–). LSc:1944–35, 1 *cw*. NYBK, 1 *cw*.
56, 57.	ER	(F&L–0). PBN:F1372, 1 *cw*. NYNS, 1 *cw*.
57a.	FR	(F&L 1648). François Randon. BxCMe, 1 *cw*.
58.	RR	(F&L–0). CyYSt, 1 *cw*.
59.	LS	(F&L–0). CtYSt, 1 *cw*.
60.	HV	(F&L–0). NYBK, 1 *cw*.
61.	W*	A0:659, on grain weights of box of *Jacobus* Francis Wolshot, 1776; on the scale: D in ○, B in ▽. BxRA: 3706, in box with label of J. F. W. in ○ and ▽. NYNS, in box of J. F. W., 1750, on ○ and ▽.
62, 63.	ADW	(F&L 1648–1650). Arthur van Dunwalt. No. 62, PCN:17196 on *Nnw* with adjuster's mark [1663]. No. 63, LSc, 1 *cw* 1648. NYNS, 1 *cw*, 1648. CtYSt, 1 *cw*, 1650.

194

64.	GDW*	(F&L 1641–1644). (See also No. 23.) Gerard van Dunwalt(?) NYNS, *cw*, 1617. BxCMe has a box inscribed: *Geeraerdt von Dunwalt . . . op den Hoeck van de oude borse in de crone tot Antwerp*. In the balance ▽ is Cologne sign No. 7.
65–70.	IFW†	J. F. Wolschot. BxRA:3706, *cw*. CtYSt, 2 *cw*. PBN, *cw*. NYNX, *cw*. NYBK, No. 69 and No. 70 on *cw* and on grain weights. BxCMe has a weight box by *Joannes Fransius* [sic] *Wolschot balans tot antwerpen 1730*. In it are 11 weights with sign No. 77.
71.		One *cw* in NGN in a Hamburg box of 1587.
72.		(See A. de Witte, 1895, Plate 2.) François de Batist.
73–77.		(From A. de Witte, 1895, Plate 2.) Notice No. 75 with J instead of I.

FRENCH MASTERSIGNS

Local monetiform and other weights, still extant in great numbers and dating from the thirteenth to the eighteenth centuries, originated from many different cities of southern France. Forien de Rochesnard and Lugan (F&L) have carefully listed the various French towns and their known scale makers (*balanciers*). We list here only those from Paris and Lyons, because these seem to have been the most important centers of manufacture in France during the seventeenth and eighteenth centuries and because the author can make some small additions to the valuable data of F&L.

---

* Another balance maker from Antwerp (F&L 1699) was Peter van Dunwalt. BxCMe has a coin and weight box of his of 1699. A box with label by Gerard van Dunnewaldt [sic] of 1677, his address given as *Hamborgh*, is in possession of SN (F&L 1641–1644). See also No. 64 of this chart. Among the balance makers van Dunwalt of Antwerp, besides Geeraerdt (1641–44) and Peeter (1696), there was also an Aertus van Dunwalt. BXCME has a box inscribed: *Aertus van Dunwalt, maeckt ende verkopt dese gout-gewichte op den hoeck van de oude. in de croone tot Antwerpen*. In the round pan of the balance is the mastersign 5a of Chart 7. This is the same address as that of Geeraerdt van Dunwalt. The mastersign in the balance of Geeraerdt's box and some similarly inscribed on gold scales in Antwerp boxes indicate that Geeraerdt van Dunwalt was for a certain time the Aichmeister of the Free City of Cologne. We do not know who was Aichmeister in Cologne between 1596 and 1606 (Kisch, 1960b, p. 67).

For other Belgian scale makers see Forien de Rochesnard and Lugan.

† There have been at least three balance makers named J. F. Wolschot in Antwerp. One was "Joannes Franscus" [sic]. WB:G 4295 has a weight box inscribed by him from 1738, and CtYSt has one labeled 1730. The other Wolschot can be dated between 1750 and 1776. He was "Jacobus Francis" (AO: 659). F&L mention also Joseph François Wolschot, 1785. A. de Witte (1895, pp. 63 ff.) mentions: (i) Jean-François Wolschot, appointed as adjuster 29.6.1749, d. 1765; (ii) Jacques-François Wolschot, son of (i), appointed as adjuster 11.4.1756; (iii) Joseph-François Wolschot, son of (ii), applied for appointment as adjuster 23.9.1785.

Paris

According to Machabey (1949, pp. 59, 60) in Paris in May 1510 a sworn balancier named Hubert Delamare and the master balance makers Millet Jean, Mathieu Galbin, and Pierre Dobry were registered. In an ordinance of October 3, 1519, eleven Paris balanciers are mentioned: Martin Gilles, Mathieu Gallebin, Hubert de la Mare, Johan Gallebin, Julien Chouet, Germain Poijret, Estienne Jouan, Jehan Baron, Pierre Mazurier, Jehan de la Mare, Jehan Choquet.

Each balancier in Paris and apparently elsewhere in France (for instance in Rouen: F&L, p. 97) was obliged to have as his personal mark a crowned letter, representing his name (Diderot's *Encyclopédie*, vol. 4, p. 253; Jaubert, p. 199). According to the labels in French weight and scale boxes this crowned letter, which was a symbol of royal privilege, seems also to have been the sign of the house in which the balancier had his workshop and business. The sign was impressed on nested weights which he sold, and which he had either made or adjusted, and it served on the labels in the weight boxes as a kind of business address (see Fig. 98). It was also used subsequently without change by other scale makers, even if they had a different initial, but this seems to have been a practice only in the eighteenth and early nineteenth centuries.

A list of the crowned-letter hallmarks of Paris balance makers (from F&L), with some of my additions, follows. The letters added in parentheses refer to the collections where I found the objects (see list of abbreviations on p. xix).

A	Cannu l'Aîné 1724, rue St. Denis; Pierre Frémin 1768. CtYSt: CIN-8. In Nuremberg nested weight (*Nnw*); (PCl:8080) 16 M.
⚜A⚜	Mourette 1768–1784; Malice after 1784. On label in weight box of LSc:1942–30: *A l'A Couronné/Rue St. Martin No. 41 vis a vis celle St.-Méry (ci-devant rue de la Ferronerie)/GODEFROID/Successeur de Madame Veuve MALICE/Md. Balancier-Ajusteur/*. The sign is also in a *Nnw* (64 M) at PCN:3262.
B	Renauld 1764. This sign is found also in a French nested weight (end of the 18th or early 19th century) with livre and gramme designation (NYBK) and in another (LScC) from the beginning of the 19th century.
C	(F&L-0). Charpentier, Cheville, NGN:WH118. On label in weight box: *Au C̆ courronné Rue St. Denis No. 73 vis-a-vis celle des Lombards au coin de la petite Place Gastine/CHARPENTIER, Balancier Ajusteur du Tresor de la Couronne & de la Banque de France.* "Tresor de la Couronne" probably

196

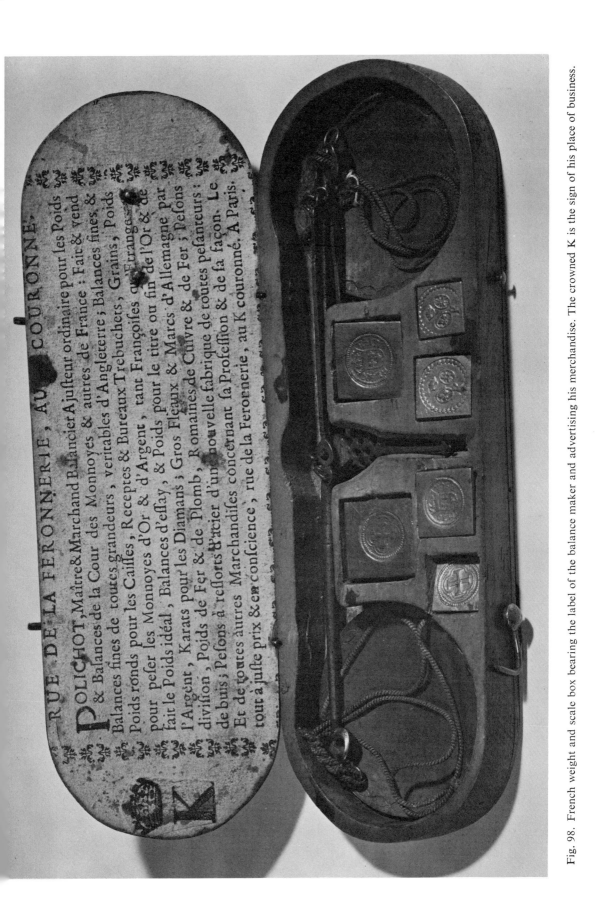

RUE DE LA FERONNERIE, AU COURONNE.

POLICHOT, Maître & Marchand Balancier Ajuſteur ordinaire pour les Poids & Balances de la Cour des Monnoyes & autres de France : Fait & vend Balances fines de toutes grandeurs, veritables d'Angleterre ; Balances fines & Poids ronds pour les Caiſſes, Receptes & Bureaux Trebuchers, Grains, Poids pour peſer les Monnoyes d'Or & d'Argent, tant Françoiſes qu'Etrangeres fait le Poids idéal, Balances d'eſſay, & Poids pour le titre ou fin de l'Or & de l'Argent, Karats pour les Diamans ; Gros Fleaux & Marcs d'Allemagne par diviſion, Poids de Fer & de Plomb, Romaines de Cuivre & de Fer ; Peſons de buis ; Peſons à reſſorts & d'acier d'une nouvelle fabrique de toutes peſanteurs : Le Et de toutes autres Marchandiſes concernant ſa Profeſſion & de ſa façon. tout à juſte prix & en conſcience, rue de la Feronnerie, au K couronné. A Paris.

Fig. 98. French weight and scale box bearing the label of the balance maker and advertising his merchandise. The crowned K is the sign of his place of business.

indicates a pre-Republican date. In the collection of Wittop Koning in Amsterdam is a Paris *wsb* with a label reading: *Au C̄ couronné Rue St. Denis 85 au coin de la rue de la Féronnerie vis-a-vis la rue Troussee-Vache Cheville, fils Maitre et M^d Balancier Ajusteur.*

D — Le Coqu 1750. Boully Parent, gendre de M. Denis 33 rue des amis. End of 18th century. PLa has a weight box (ca. 1800) with label: *Au D couronné/Boully/Balancier Rue des Arcis,/ vis-a-vis la Signe vert No. 19.*

G — Girard, 54 rue St. Martin.

I — Cheville père, rue St. Denis. PCN:17572 has this sign in a *Nnw*.

K — Tilly 1772. Polichot, rue de la Féronnerie, Fournier. PBN possesses a *wsb* with the label: *Rue de la Feronnérie au K couronné. /Polichot, Maitre & Marchand Balancier Ajusteur...*

A Paris (see Fig. 98). NYBK, on label in *wsb*: *Au K couronné/ Rue de la Feronnérie la septième Boutique/en entrant par la Rue St. Honoré No. 12/A PARIS/FOURNIER/Balancier, Ajusteur Verificateur/tient Fabrique de Poids et Balances.* A *wsb* with identical label has KnF. The nest in this box with a crowned P has gram denomination and must therefore be ca. 1800 (see crowned P). PBN has a box by: *Polichot, Rue de la Feronnerie au K couronné.* NYBK has another box without label; on the cups of a nested weight is impressed the crowned K.

L — Fourché 1777, rue de la Feronnerie, François le Goix. NYBK, *wsb*, apparently early 19th century; on its label: *François le Goix Md Balancier, adjousteur des poids d'or & d'argent... demeure rue S. Denis, proche les Innocens à l'L couronnée a Paris.* NYBK has also a nested weight of a peculiar (not Nuremberg) shape inscribed: *500 GRAMMES*, with a crowned L. PLa has a weight box with the label: *L à Paris 1700. S. Denis proche des Innocens a L couronné attenant la rue de Ferronerie.*

M — Malice 1777–1784.

N — François Trifon 1719, rue de la Fernaneria [sic] (F&L).

P — Pilloy, 73 rue St. Denis 1777; later, Pourin and also Charpentier.

ZMI has a *wsb* with label: *Au P̂ couronné Rue St. Denis No. 78 du cote et pres la rue des Lo[mb]ards a Paris/ POURIN/.* The box contains two nested weights, one 200 gm, the other 1 M. PLu, MRB:8777 BxRA, and CtYSt:B 65 each have one *wsb*, each with the following label: *AU P̂ couronné /Rue St. Denis No. 73 vis-a-vis celle des Lombards/au coin de la Petit Place Gastine/CHARPENTIER/* . . . PCN:17191, label in *wsb*: *AU P̂ couronné Rue Denis . . . Pourin successeur de Pillois Paris.* In a box in PLu and one in PLa the address of Pourin, "Successeur de Pillois" is given: *Rue Denis* [sic] *No. 38 entre la Rue Trous Vache et celle des Lombards à Paris.*

Q̂ Canu 1748–1760; Barbin 1760–68; Chemin 1768–; Fourché Chemin rue de la Ferronnérie. CtYSt:B 63 possesses a *wsb*, end of 18th century with label: *AU Q̂ couronné/CHEMIN/Mbre de l'Athénée des Arts/Balancier Mécanicien rue de la Férronnerie No. 4 à Paris.* Similar labels in two weight boxes are owned by PLa. NYBK:B 26 has a *wsb*, early 19th century, of unusual form, on the label: *AU Q̂ couronné No. 4/Fourché-Chemin./ Balancier Mécanicien, Rue de la Feronnerie, a Paris.* Guillemot *et al.* (1902, p. 447) mention this Fourché as "gendre et successeaur du citoyen Chemin." The Q̂ is also on a *Nnw* in CtYSt:ClN 17 and in PCN:1790, 17193, and 4257.

R̂ (F&L–0). PCl on a *Nnw*.

T̂ Germain des Vignes 1772. Pourin. Guillemot et al. (p. 447) mention this sign as: *Rues Denis entre la Rue Trousse-Vache et celle des Lombards Pourin successeur de Pillois.*

Ŵ Pierre-Charles de Villiers 1767.

It would require a special study to clarify the use of crowned letters as business addresses by weight and scale makers in Paris. Expressions like *AU Q couronné* (at the crowned Q) or *AU Q couronne, No. 4 rue Feronnerie*, confirm the opinion that the crowned initial is the insigne of a certain house, for it is known that the balancier Chemin had his workshop in No. 4 rue de la Feronnérie.

In spite of the legal obligation, apparently not all Paris balanciers used a crowned letter in the eighteenth and nineteenth centuries. An elaborate weight box (KnZ:HM-188) is marked: *Poids pour Essais* [standard weights] *Empère*

199

à Paris. One nineteenth-century weight box (CtYSt:B-33) has impressed in the lid *Le Carat de 205 millig/L. Empere 71/Rue de Turbigo 71./Paris.*

Rouen

According to Forien and Lugan (p. 97) the letters B and G with a crown were also used by balanciers in Rouen. This may have been an adjuster's mark, since it was used in Lyons and elsewhere, and no proof is given that it was ever used as trademark or house insigne. To the list of masters from Rouen given by F&L may be added a box at NYNS with the label: *On trouvera chez Hue, M^d Balancier rue de lave Conte/N° 207 a Rouen...* In Rouen also each balance maker was supposed to have a mastersign (Balance-makers' Statute of 1415). See Ouin-Lacrois (1850, p. 467). He mentions (p. 197) that in the eighteenth century there were six or seven master balance makers active in Rouen; their number was still the same in 1850 (p. 375).

Bordeaux

In discussing inscriptions on weights and scales I should like to add to F&L's list two other balance makers of Bordeaux. SMk has a weight box inscribed *Rafiné par Jean Reyne mettre Rafineur dans la rue de La roncelle a bourdeau le 17 may 1709*. His sign is IR under a crown. CtYSt has a weight and scale box with the label: *A. Fournel/Balancier du Roi/Marche Royal/No 5/ A Bordeaux*.

Lyons

A few words concerning Lyons, the most important center of weight making in France, are in order. Again, F&L's knowledge, gained from archival documents and from weight and scale boxes, is especially complete and valuable. The following data contain a few additions to their list.

MC   (F&L 1664–1668). PBN:33 has a *wsb* by Mathieu Clot, rue Tupin 1662. His sign MC and that of his father Noe Clot (C) are on most of the weights.

IB   (F&L 1668–1726). Jacques Blanc. StAD:43457 has a weight box inscribed: *Jacques Blanc Rue tupein* [sic] *au 3 Dophny* [?] *a Lion 1666.*

IP   This sign surrounded by the name *CHAUDET* to be found in each pan of a scale in a box in NYBK inscribed: *Jean pierre Chaudet/Rue tupin a lion 1676.*

♔
ID   See page 202.

♔
LF   (F&L 1648–1668). Andre Le Fran. PCl has a weight box
⚜   inscribed: *P. Rogon [A]nd[r]é le fran Rue tupin a lion 1614.* In
LF   center of both pans of balance LF surrounded by *LE FRAN.*
In the money weights the sign LF is in a weight box in PBN
inscribed: *André le Fran/Rue tupin/ . . . A lion/1651.* NYBK
and BaW have boxes inscribed: *André le Fran . . . Rue tupin
♔   a lion 1611.*
LF   NYBK, *wsb* inscribed: *Andre le fran/Rue tupin a lion 1660.*
✕
♔
This sign is in both pans of the scale. On the weights: LF.
The complete sign is also in both pans of the balance of
ZL:LM 23044 in a *wsb* inscribed: *ANDRE le Fran/Rue
Tupin/a lion/1641.* On the weights is the sign LF.
⚜   PCN:6226 and BsH:1949–70 have this sign on weights in box
♔   inscribed: *A Lefran 1652 Lion.* Considering the time span
LF   (1611–1668) one wonders whether André le Fran in rue Tupin
in Lyons would have been both father and son, although it is
improbable, but of course not impossible, that a master could
♔   have been active for 57 years.
PG   Listed by F&L, p. 95, as Paul Garcin 1668. PCl has a *wsb*
inscribed: *Pierre Grand a La Rue des Chapeliers proche La Rue
♔   tupin a Lyon 1661.* This sign is in both pans of the balance.
LG   Laurent Grosset 1679–1723.
(or un-
  crowned)

♔
IL   (F&L 1659–1661, rue Mercière). Jacques La Lune. BsH:1942–
24 has a *wsb* inscribed: *Jacques La Lune a Lyon Rue Merciére
♔   1677.*
CM   (F&L 1668–1676, rue des quatres Chapeaux). Claude Marion.
CoUl has a *wsb* inscribed: *Claude Marion mte balancier/De-
meuren a la grenette vis a vis/Lagrand Rue a Lyon/1678.* In both
pans of the balance is an uncrowned rampant lion flanked by
the letters C and M.

IM   (F&L, p. 94, 1668–1671). Jacques Michaud. ZL has a *wsb*
inscribed: *Jacques Michaud en Rue/Mercier a Conseigne P./
Michel a Lyon 1663.*

IP	Joseph Pascal. NYBK, in *wsb*, 18th century, burned into lid: *IOSEPH + PASCAL/RUE♡D♡4♡CHAPAVX♡A♡LYON*; no signs on the weights; original balance missing. WBG:3774 and ZL:EM-14542 have similar boxes with the same inscription but without the year. PBN:33 has a *cw* in a Lyons box dated 1662 which has IP with *PARIS* beneath it. NYBK, on *cw*.
♔ IP	(F&L, p. 94, 1676–1702, rue Tupin). Jean Pingard. NGN:31 has a *wsb* inscribed: *Jean Pingard a la grande/Rue aux dauphins a Lyon 1726*.
♔ (ID)	On one weight of the box described above is ID. This sign appears also in both pans of the scale and beneath it a U. The same sign is in pans of a scale in LWH:173958, by Pingard, rue Tupin 1684.
♔ ℞	Pierre-Louis Rogon (1668–1676). Repeatedly, instead of his initials PR, I found a monogram in which P and L (for Pierre-Louis) had a common stem; e.g. in LWH:1949–70, in a box inscribed: ℞ *Rogon [A]ndré le fran Rue tupin a lion 1669*.
FD	PBN; this sign is on some weights in a box of Dominique Pascal; also in PBN:Y 21587.
IG	PBN:YZ1524 has a *wsb* with *IEAN GROSSET P. VETURIN A LION* burned into the lid.
[cock]	(F&L-0). SMk has a *wsb* by: *Thomas Lecocque Rue . . . Tupin a Lyon 1738*. This sign is in each pan of the scale.

In addition to the typical marks of the Lyons sealers the following marks have been found on objects in the collections noted. They are not found in F&L:

♔ FG	NYNS.
♔ ID	NH:31.
♔ P	NYBK.
NR (uncrowned)	CR.
♔ GD	NYBK, *cw*.
M✗D	CtYSt; *cw*.
C F	Between C and F are a bird and a crown. PBN:F 1408, on both pans of French collapsible scale.
⚜	
IM	(F&L without crown: Jacques Michaud 1668–1671). CtYSt has a *cw* with this sign.

P	NYBK, on *cw* in a box of Claudet 1676. CtYSt, on a *cw*, also in relief under a crowned porcupine on three *cw*.
♛ LP	(F&L–0). PCl, this sign is in each pan of scale in a box of Le Fran 1614.
♛ R	CtYSt, on *cw* (Lyons).
♛ CR	CtYSt, on *cw* (Lyons).
♛ FR	CtYSt, on *cw*.
♛ P.R	CtYSt, in cup of *Nnw*.
S (uncrowned)	CtYSt, on *cw* (Lyons).
 ♛ ⚜ S	CtYSt, on coin weight (Lyons).
 ⚜S	CtYSt, on coin weight.
 ♛ ⚜ X	CtYSt, on coin weight.
C[fish(?)]V	PCN:17186. In both pans of a scale. The sign between C and V seems to be a fish.

MASTERSIGNS OF OTHER COUNTRIES

It would be worthwhile to follow the history of weights and balances in all countries, but that would be far too great a task for a single author. Fortunately, some studies similar to this one have been made for parts of Europe (see the Bibliographical Notes). Perhaps this present monograph will spur young scholars to undertake an extensive study of the history of weights and scales in other countries, especially in England, Italy, Russia, and Spain. Archaeologists could find a valuable field of study in the weights and scales of ancient populations and their portrayal in ancient art.

GRAPHIC SYMBOLS OF WEIGHT UNITS

In addition to understanding the symbols inscribed by masters and sealers on weights and scales, the scholar must become familiar with the symbolic abbreviations used in ancient and modern literature for the various weight units. Those pertaining to the metric-decimal system have already been mentioned in Table 1, but the abbreviations of weight units from antiquity are more intricate and have been largely forgotten. Those appearing on Roman and Byzantine weights have been assembled and published by Pink (1938) and are reproduced here in Table 4.

No less important are the symbols used by ancient scribes, which were adopted by medieval writers and then by the printers after the fifteenth century. These symbols had a practical importance, especially in medicine, because of their continuing dependence on Greek and Roman tradition. The symbols in prescriptions found in old manuscripts had to be understood to be copied and used by later generations. Charts 8 and 9 have been reproduced here to aid physicians, pharmacists, and antiquarians in their perusal of manuscripts dating from the age of humanism, which fostered great interest in all relics from Greek, Latin, and Hebrew antiquity.

Chart 10 is a remarkable table taken from Bernard's book, *De Mensuris et Ponderibus Antiquis* (1688). It contains the symbols for the standard weights and measures of China in the seventeenth century and the symbols for the numerals from 1 to 46. At the lower left is a table of the standard units, which is self-explanatory; the name of these standards with the author's Latin explanation is at the upper left.

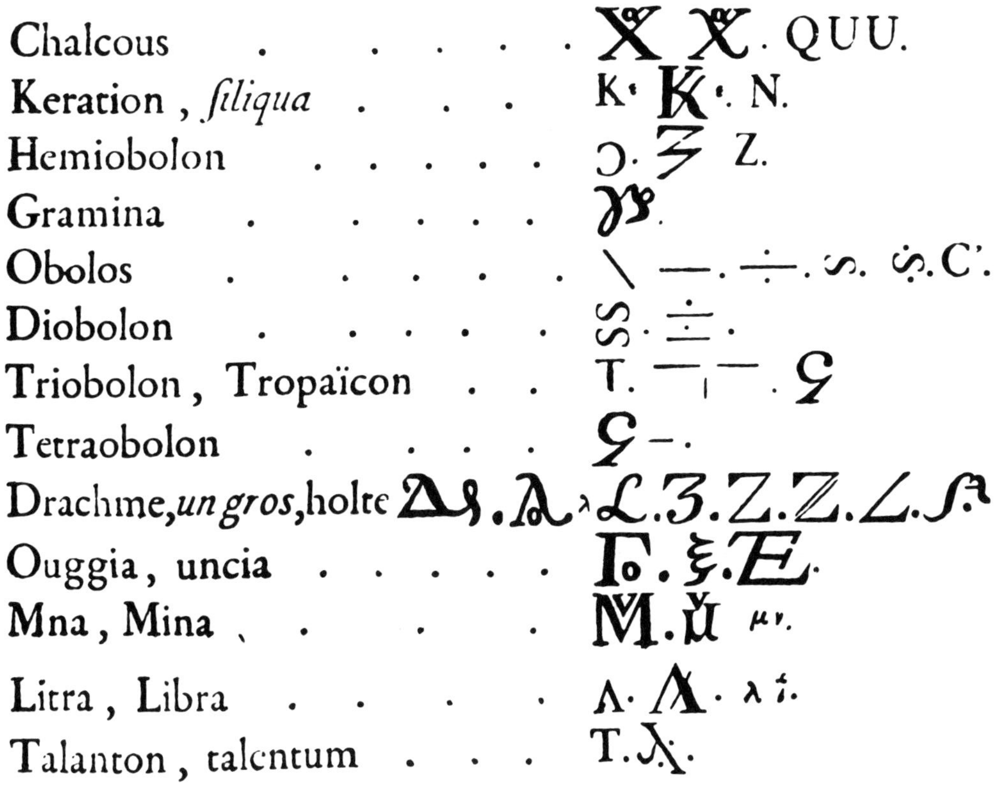

Chalcous	Ҳ Ҳ . QUU.
Keration , *siliqua* . . .	K· **K** ·. N.
Hemiobolon	ɔ . Ȝ Z.
Gramina . . .	⅄Ᏸ.
Obolos	\ — . ÷ . �568 . �. C·.
Diobolon	ss ÷ .
Triobolon , Tropaïcon . .	T. — — . Ϛ
Tetraobolon	Ϛ — .
Drachme, *un gros,* holte	Δᏸ . ⅄ Ꮑ ℒ . ℨ . Ζ . Ζ . ⅃ . Ꞙᴿ
Ouggia, uncia	Ᏻ . ξ . E .
Mna , Mina , . . .	M . ꙟ μ̌ν.
Litra , Libra	⋀ . Ⲗ . λ ἴ.
Talanton , talentum . . .	T . λ .

Chart 8. Symbols for Greek weights.

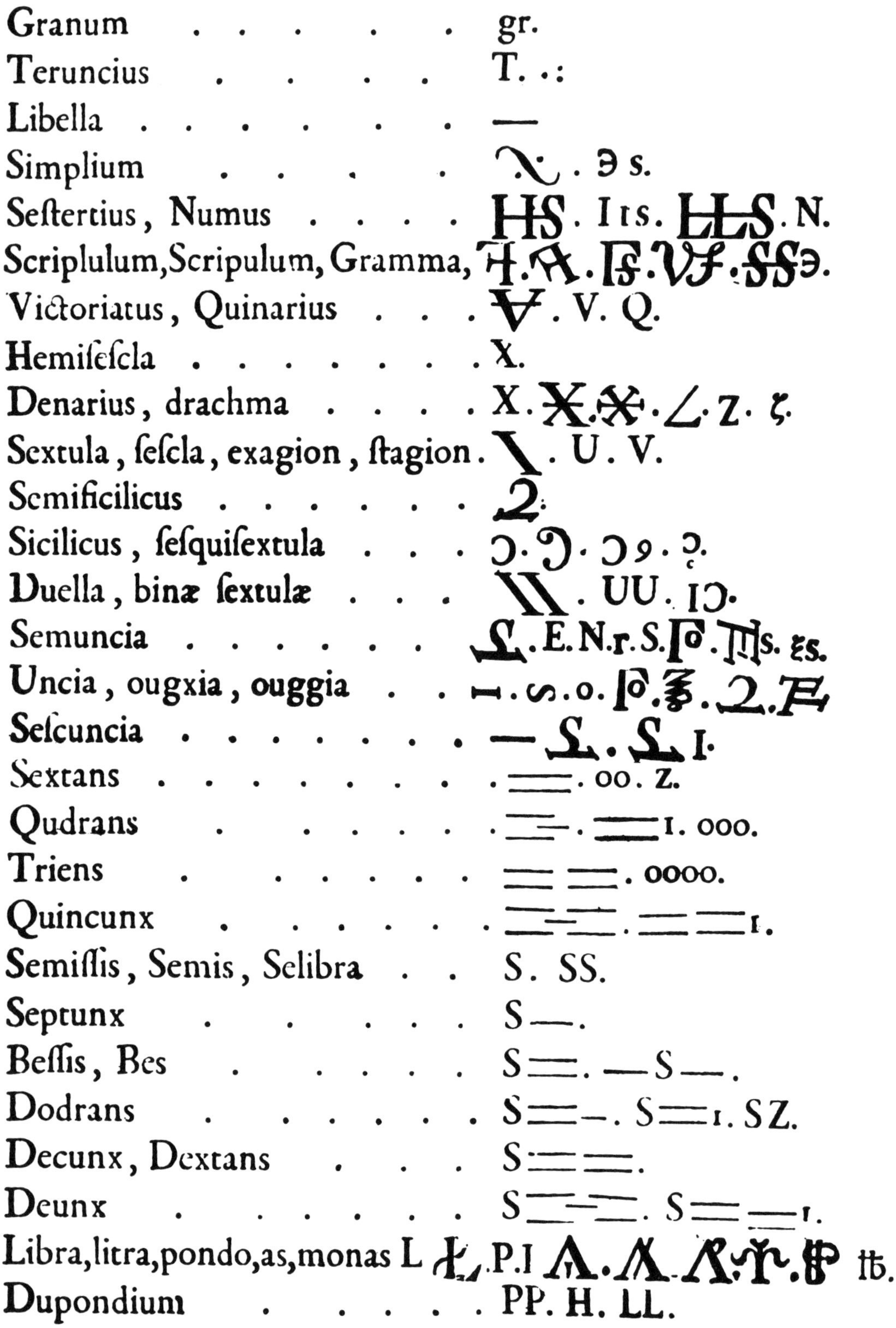

Granum	gr.
Teruncius	T. .:
Libella	—
Simplium	⅄ . Ɜ s.
Sestertius, Numus	HS . Iɩs. LLS . N.
Scriplulum, Scripulum, Gramma,	₸ . ⅄ . Iʂ . Vʃ . SSƏ.
Victoriatus, Quinarius . . .	V . V. Q.
Hemiſeſcla	X.
Denarius, drachma	X . X . X . ⟋ . Z . ʒ.
Sextula, ſeſcla, exagion, ſtagion .	⅄ . U . V.
Semificilicus	Ƨ
Sicilicus, ſeſquiſextula . . .	Ɔ . Ɔ . ƆƏ . Ɔ.
Duella, binæ ſextulæ . . .	Ⅱ . UU . CⅠ.
Semuncia	Ƹ . E . N . Γ . S . Ƨ . Ⅱs . ℥s.
Uncia, ougxia, ouggia . . .	⊢ . ω . o . Ƨ . ℥ . Ꝛ . ℈
Seſcuncia	— Ƹ . Ƹ Ⅰ.
Sextans	═ . oo . Z.
Qudrans	═ . ═Ⅰ . ooo.
Triens	═ ═ . oooo.
Quincunx	═ ═ . ═ ═Ⅰ.
Semiſſis, Semis, Selibra . .	S . SS.
Septunx	S —.
Beſſis, Bes	S ═ . — S —.
Dodrans	S ═ — . S ═Ⅰ . SZ.
Decunx, Dextans . . .	S ═ ═.
Deunx	S ═ ═ . S ═ ═Ⅰ.
Libra, litra, pondo, as, monas L	Ⱡ . P . I . Λ . Λ . Υ . Υ . ₧ ℔.
Dupondium	PP . H . LL.

Chart 9. Symbols for Roman weights.

Chifres Arabes.	Chifres Grecs.		Chifres Romains.
I	$\acute{\alpha}$	I	I.
2	β'	II	II.
3	γ'	III	III.
4	δ'	IIII.	IV.
5	$\acute{\epsilon}$	Π. Γ.	V.
6	ς'. ⟨ᶾ⟩	Π I	VI.
7	ζ'	Π I I	VII.
8	$\acute{\eta}$	Π III	VIII.
9	ϑ'	Π IIII	IX.
10	$\grave{\iota}$	Δ	X. X̶.

Chart 9 continued.

Chiffres Arabes.		Chiffres Grecs.	Chiffres Romains,
20	χ'	ΔΔ	XX.
30	λ'	ΔΔΔ	XXX.
40	μ'	ΔΔΔΔ	XL.
50	ν'	⊡.⊡	L.
60	ξ'	⊡ Δ	LX.
70	ο'	⊡ ΔΔ	LXX.
80	π'	⊡ ΔΔΔ	LXXX.
90	Ϟ'.Ϟ	⊡ ΔΔΔΔ	XC.
100	ρ'	H. Ⱶ. Ⱶ. E.	C.
200	σ'	HH	CC.
300	τ'	HHH	CCC.
400	υ'	HHHH	CCCC.
500	φ'	⊞.⊞	IↃ. D. Ð.
600	χ'	⊞H	DC.
700	ψ'	⊞HH	DCC.
800	ω'	⊞HHHH	DCCC.
900	Ϡ'.πι.	⊞HHHH	CM.
1000	α	X	M. CIↃ. Ī. ∞. ⅭⅠↃ. ⅄.
2000	β	XX	MM.
3000	γ	XXX	MMM.
4000	δ	XXXX	MMMM.
5000	ε	⋈.⋈	IↃↃ. Ð. V̄.
6000	ϛ	⋈X	IↃↃM.
7000	ζ	⋈XX	IↃↃMM.
8000	η	⋈XXX	IↃↃMMM.

Chart 9 continued.

Chifres Arabes.	Chifres Grecs.	Chifres Romains.
9000	θ	|X|XXXX IↃↃMMMM.
10000	,α . ä	M. M̄
20000	κ . β	M̄M / X̄X̄.
30000	λ . γ	M̄MM / X̄X̄X̄.
40000	μ . δ	M̄MMM / X̄L̄.
50000	ν . ε	M̄.M̄ ICCI / L̄.
60000	ξ . σ	M̄M / L̄X̄.
70000	ζ . ο	M̄MM / L̄X̄X̄.
80000	π . υ	M̄MMM / L̄X̄X̄X̄.
90000	ϟ . ∽	M̄MMMM / X̄C̄.
100000	ρ . ι	△ / CCCIↃↃ.
1000000	ρ̄ . ρ	H̄M̄ X̄X̄ / CCCCIↃↃↃ.
10000000	∞	Ψ.

Chart 9 continued.

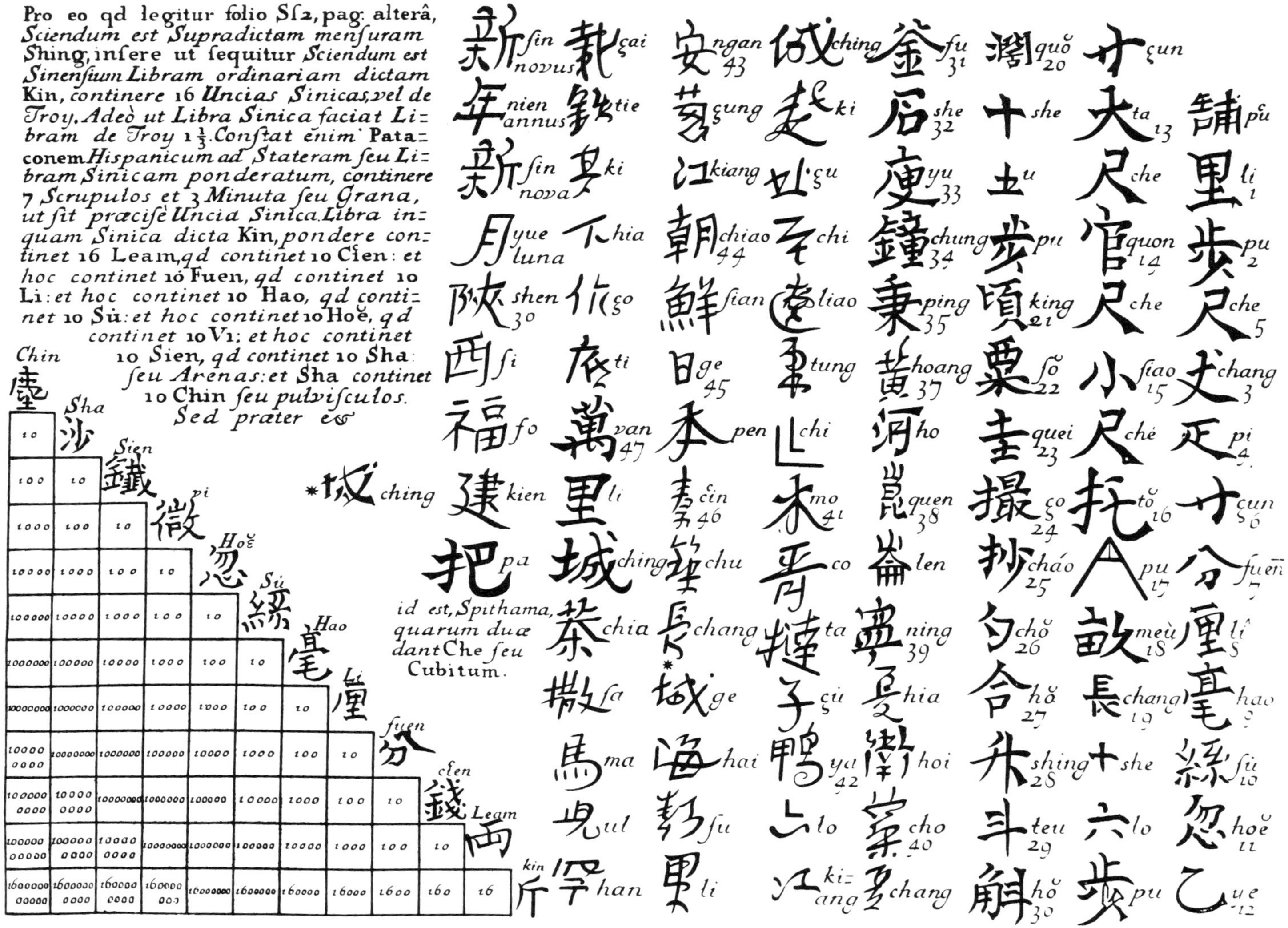

Pro eo qd legitur folio Sſ2, pag: alterâ, Sciendum est Supradictam menſuram Shing, inſere ut ſequitur Sciendum est Sinenſium Libram ordinariam dictam Kin, continere 16 Uncias Sinicas, vel de Troy. Adeò ut Libra Sinica faciat Libram de Troy 1⅓. Conſtat enim Pataconem Hispanicum ad Stateram ſeu Libram Sinicam ponderatum, continere 7 Scrupulos et 3 Minuta ſeu Grana, ut ſit præciſè Uncia Sinica. Libra inquam Sinica dicta Kin, pondere continet 16 Leam, qd continet 10 Cſen: et hoc continet 10 Fuen, qd continet 10 Li: et hoc continet 10 Hao, qd continet 10 Sù: et hoc continet 10 Hoĕ, qd continet 10 Vı: et hoc continet 10 Sien, qd continet 10 Sha: ſeu Arenas: et Sha continet 10 Chin ſeu pulviſculos. Sed præter &c
id est, Spithama, quarum duæ dant Che ſeu Cubitum.

Chart 10. Symbols for standard weights and measures in seventeenth-century China.

7. CONCLUDING REMARKS

The manufacture and use of weights and scales have been supervised by government since remote antiquity. The legal history of weighing-tools in each country would be a valuable study which is largely yet to be done. After money, weights and scales are the most important instruments of national and international commerce, and in most countries only carefully selected and certified persons were permitted by law to make them.

The guild statutes (German: *Zunftbriefe*; French: *statuts des balanciers*) have existed since medieval times and had their beginning as far back as the Near Eastern civilizations. The privileges and duties of the masters and their apprentices were established in the guilds down to the smallest details, and special laws dealt with the manufacturing process (in France, Testut, pp. 166 ff.; in Cologne, Kisch, 1960b; in Italy, *Istruzione*, 1750). The high officers of the guilds had the right and duty to enforce all laws regulating the advancement of apprentices and the manufacture of merchandise. In France, the balanciers' guild was under the jurisdiction of the board of the mint (Cour des Monies), who also graduated the masters (Diderot, *Encyclopédie*, vol. 4, p. 253). The finished product had to be checked by a government sealer. After the instrument had been purchased by a private citizen, it was still subject to periodic rechecking and re-marking by government officials to discourage cheating.

In France on February 1, 1312, Philippe le Bel ordered that all weights and scales be checked at least two or three times each year. By 1484 it had been decided by royal decree that pharmacists should check weights and scales of merchants selling sugar, wool, drugs, and spices, but only the mint was empowered to check those of the goldsmith. This law was confirmed in 1603. In miniature paintings in medical manuscripts of the thirteenth century (Venice, Salerno) a balance is shown as part of the inventory of the pharmacy (Blaxland and Bligh, 1931). The guild of the pharmacists (confirmed in 1659) had in their coat of arms a balance and the device: *"Lances et pondera servant"* (Guillemot et al., 1902 p. 177: Felibien, 1725, vol. 2, p. 928). *Lances* here means the pans or *pars pro toto*, also the scale. Wallis (1670, vol. 1, p. 67) explains the meaning of the word balance as follows: *"Bilanx a binis lancibus nomen habet."*

All countries of Europe had strict laws against the use of improper weights and scales. Whenever the government decided to change the legal standard of its weight units, the destruction of items no longer valid was imposed on their owners, and wherever such units were discovered to be still in use they

211

were confiscated and destroyed. This was the case, for example, in the Napoleonic era, when France introduced the metric-decimal system by force in Cologne and the Rhineland (Kisch, 1960b).

Innumerable laws were instituted, enforced, and subsequently changed or annulled in every civilized country to keep weighing and measuring under strict governmental control. Attention can be directed to the few publications on the subject; de Witte (1895), who was interested mainly in Belgium, published various papers in the *Revue Belge de Numismatique* (see Bibliography). Knowledge of French laws relating to weights and scales may be gained from Felibien (1725) and a real treasury of knowledge from the important books of Ouin-Lacroix (1850), Testut (1946), and Machabey (1949); and for the Netherlands from the oft-quoted book of Zevenboom and Wittop Koning. A monograph on Cologne's laws and regulations concerning weights and scales has recently (1960b) been published by the present author. These, of course, are only a beginning toward an understanding of the role of governments in the manufacture and use of weights and scales.

This would seem the place to say something about the men who made the early weights and scales, at least those in the European countries where documents and reports are still extant. A full discussion, however, would add unduly to the length of the book, and the subject is not directly pertinent. The reader is referred to thorough studies by Ouin-Lacroix (1850), Clémenceau (1909), Testut (1946), and Machabey (1949) for France; and to Kisch (1960b) for Cologne. A complete study remains for a future scholar of the history of culture, but a few aspects can be mentioned here. As in Paris, Rouen, and Lyons in France, and Cologne and Nuremberg in Germany, the weight and balance makers elsewhere were also united in a kind of brotherhood, or their group was at least incorporated into one of the big guilds of the town. In Cologne they were attached as an individual guild to the blacksmiths (*Schmiedeamt*), in Nuremberg to the coppersmiths (*Rothschmiede*). They preferred certain streets for their business, which were not always the same in different centuries. In Cologne it was mainly on the Uffm Brand in the seventeenth century and on Unter Taschenmacher in the eighteenth century. Jaubert (1773, p. 201) informs us that the patron saint of the weight and scale makers in Paris was St. Michael and their preferred church St. Innocents. They lived and conducted their business and workshops near this church, and legends on extant weight and scale boxes of the eighteenth century prove that the rue St. Denis and the nearby rue de la Feronnerie were their principal locations. In the sixteenth century le rue de la Feronnerie was *un des passages les plus frequentes de toute de la ville de Paris* (Felibien, 1725, vol. 2, p. 1044). In Lyons, at least in the eighteenth century, the rue Tupin

212

and rue des Quatres Chapeaux must have been the favorite streets of the balanciers of this city.

The products of the balance makers were, of course, variously esteemed. In Germany, weights and scales from Cologne and Nuremberg were valued most highly. Nuremberg's nested weights were so fine as to give this city a monopoly on their manufacture. In eighteenth-century France, according to Jaubert (1773, p. 193), fine balances (*trébuchets* or *balances d'essai*) came from Lyons and from Forez, but those made in Paris were considered best.

The weight maker had his individual mark which, as a rule, he was compelled by law to emboss on the products of his workshop; this was true in Cologne, Nuremberg, Paris, and probably elsewhere. In Nuremberg and in Paris (Jaubert, p. 200) this had also to be embossed on a piece of metal which was kept by the government; it thus became part of a list of registered trademarks. One of the Nuremberg registers is preserved (see Stengel, 1915), but Cologne's have not survived. We know, however, that in Cologne the masters of the *Blechschläger's* guild (sheet-metal workers) were also obliged to register their mastersigns on a tin plate and deposit them (Kisch, 1960b, p. 166) with the appointed *Commissaries*.

In the Gruuthuse Museum in Bruegge (Holland) I found in a closed cabinet various brass and tin plates with embossed or engraved names and dates, probably of officers to check weights or of weight makers. They would be very important to this topic, but up to the present time I have not been able to get photocopies of them for further research.

We are well informed about the rules in France through Jaubert (pp. 200 ff.), Testut, and Machabey. Each weight and scale maker was supposed to have his individual mark assigned to him by "*les Jurés.*" As a rule it was the initial letter of his name. This sign (as in Nuremberg) had to be embossed on a copper plate and deposited in the government archive (*au greffe de la Cour des Monnoies*) at the mint. The initial was surmounted by an open crown (*une couronne fleurdelisée*). The mark of the mint itself according to Jaubert (p. 200) was a simple, embossed fleur-de-lis.

In Paris the early balanciers admitted only natives of Paris to their union. Their constitution of 1325 had been renewed in 1691 and 1695 (Diderot, 1778, p. 253). Their apprenticeship lasted five years and had to be followed by two years of working with a master; in Cologne the apprenticeship lasted six years.

In Europe since the nineteenth century, the personal art of the weight and scale maker has progressively been displaced by the mass production of factories and the rules of international convenience, but for those who retain a feeling and a certain love for the personal touch in the objects of daily life, the study of ancient weights and scales offers the reward of the joy one may gain from any object of art.

APPENDICES

APPENDIX 1

Ancient weight standards and their values in grams (gm) and kilograms (kg) according to Alberti (1957, pp. 24 ff.), Berriman (1953), Weigall (1908), and others.

1. Babylonian Weights

In Lagash in the third millennium (ca. 2350 B.C.) the mina was 477 gm; 400 years later in Lagash 120 minas = 2 biltu = 60.54 kg. Under Salmanassar V of Assyria (726–722) the norm was

$$1 \text{ heavy mina} = 1004.72 \text{ or } 1008.45 \text{ gm}$$
$$1 \text{ light mina} = 502 \text{ gm}$$

Under Nebuchadnezzar II (605–562 B.C.) the mina = 489.15 gm, equal to that of the Third Dynasty of Ur (ca. 2100 B.C.).

At the time of the Amarna (14th century B.C.) 30 kg = 1 biltu = 60 manus.

$$1 \text{ manu} = 60 \text{ shiglu} = 500 \text{ gm}$$
$$1 \text{ shiglu} = 180 \text{ se'u} = 8.3 \text{ gm}$$
$$1 \text{ se'u} = 0.046 \text{ gm}$$

According to Thureau-Daugin (1938, p. xiv), Babylonian weights in the time of the First Dynasty:

Sumerian name	Akkadian name	Equivalent
še	šeu	1/80 sicle
gin	šiklu	1/60 mine
mana	manu	1/60 talent
gun	bilto	60 mines

2. Egyptian Weights*

(The basic system: 1 sep = 10 debens = 100 kites.)

Old Kingdom (ca. 2900–2540 B.C., 3rd–10th Dynasties)
Egyptian gold deben or stater standard = 12–14.2 gm
Middle Kingdom (2160–1785 B.C., 11th–17th Dynasties)
  (a) Egyptian gold deben or stater standard = 12–14.2 gm
  (*b*) Phoenician gold deben or stater standard = 14.6–15.9 gm

---

* According to Weigall (1908) and Berriman (1953).

New Kingdom (1580–1100 B.C., 18th–20th Dynasties)
 (*a*) Egyptian gold deben or stater standard = 12–13.8 gm
 (*b*) Phoenician gold deben or stater standard = 14–15.9 gm
 (*c*) Egyptian kedet standard = 8.8–10 gm
Late Kingdom
 (*a*) Egyptian deben or stater standard = 11.5–13.5 gm
 (*b*) Phoenician deben or stater standard = 14–15.5 gm
 (*c*) Egyptian kedet standard = 8–9.5 gm
 (*d*) Alexandrian stater standard = 10–10.95 gm

The dates assigned to various periods of antiquity have undergone changes in recent years, and without much more study of extant objects and documents a definite opinion concerning standards in Egypt at a definite time is not feasible.

The kedet standard (also called kite or kat) was probably introduced between the Thirteenth and Eighteenth dynasties. Its origin is not known and it is also found (8.8–10 gm) in Syria and Troy, one deben being equal to 10 kedets (Weigall 1908, p. ix). According to Weigall, 1/12 deben is regarded as equivalent to the Assyrian shekel and is called "a piece"; with the diminishing value of the deben, 1/10 of it was 1 kedet or a piece.

3. Jewish Weights*

kikar	1				
mana	60	1			
shekel	3,000	50	1		
beka	6,000	100	2	1	
gera	60,000	1,000	20	10	1

According to the Prophet Jechezkiel they were:

kikar	1				
mana	60	1			
schekel	3,600	60	1		
beka	7,200	120	2	1	
gera	72,000	1,200	20	10	1

The numerical inscriptions on ancient Jewish weights as deciphered by Yadin (1960) read as follows:

I	II	III	٦	Λ	(I⩘)	(II⩘)	Τ	٦Λ	—	(=)	≡
1	2	3	4	5	6	7	8	9	10	20	30

---

* According to the careful studies of A. S–N. (E. Stern) (1909, p. 864).

The figures were actually found on weights and verified by Yadin. Those in parentheses are interpolated (see Fig. 42). The sign ⴲ or ⵝ or ⵠ indicates "shekel."

Some weights (Fig. 42) bear the inscription 𐤉𐤐 (Pim), others 𐤍𐤎𐤐 (Nesef).
The average weight of individual weights found in Israel is:

$$\text{shekel} = 11.39 \text{ gm}$$
$$\text{nesef} = 9.840 \text{ gm}$$
$$\text{pim} = 7.808 \text{ gm}$$
$$\text{beka} = 6.112 \text{ gm}$$

3a. Jewish Weights*

$$1 \text{ kikar} = 60 \text{ minas} = 49.11 \text{ kg}$$
$$1 \text{ mina} = 50 \text{ shekels} = 0.8185 \text{ kg}$$
$$1 \text{ shekel} = 20 \text{ gera} = 16.37 \text{ gm}$$
$$1 \text{ shekel hamelech} = 14.55 \text{ gm}$$
$$(\text{King's shekel})$$
$$1 \text{ beka} = \tfrac{1}{2} \text{ shekel} = 8.185 \text{ gm}$$
$$1 \text{ gera} = 1/20 \text{ shekel} = 0.818 \text{ gm}$$

4. Greek Weights†

$$1 \text{ talent } (\tau\acute{\alpha}\lambda\alpha\nu\tau\sigma\nu) = 60 \text{ minas } (\mu\nu\alpha\grave{\iota}) = 26.196 \text{ kg}$$
$$1 \text{ mina } (\mu\nu\alpha) = 100 \text{ drachmas } (\delta\rho\alpha\chi\mu\alpha\grave{\iota}) = 435.6 \text{ gm}$$
$$1 \text{ drachma } (\delta\rho\alpha\chi\mu\acute{\eta}) = 6 \text{ obols } (\acute{\sigma}\beta\sigma\lambda\sigma\iota) = 4.366 \text{ gm}$$
$$1 \text{ obolus } (\acute{\sigma}\beta\sigma\lambda\sigma s) = 2 \text{ hemiobols } (\acute{\eta}\mu\iota\sigma\beta\acute{\sigma}\lambda\sigma\iota)$$
$$= 8 \text{ chalkoi } (\chi\alpha\lambda\kappa\sigma\iota) = 0.728 \text{ gm}$$
$$1 \text{ chalkous } (\chi\alpha\lambda\kappa\sigma\hat{\upsilon}s) = 0.091 \text{ gm}$$

Alberti (p. 38) believes this table represents the Attic talent, but the talent from Aegina = 45 kg, and the Syrian and Ptolemaic talent = only 7 kg.

5. Roman Weights

Roman weights have been carefully studied over the years, and their values and subdivisions are well summarized in Table 11 from Hultsch, in Table 12 from Pink, and in Table 13 from Pauly. The values in grams in Tables 11 and 12 have been added by the present author.

---

* According to Alberti (1957, p. 29).
† According to Hultsch (1862, Table 12).

TABLE 11. Standards of Roman Weights

	Grams
1 siliqua	0.189
1 obolus = 3 siliquae = 1 dimidium scripulum	0.568
1 scripulum = 2 oboli = 6 siliquae	1.137
1 dimidia sextula = 2 scripula = 4 oboli	2.274
1 drachma = 3 scripula = 6 oboli = 18 siliquae	3.411
1 sextula = 4 scripula = 8 oboli	4.548
1 sicilicus = 6 scripula = 2 drachmas	6.822
1 semuncia = 2 sicilici = 4 drachmas	13.644
1 uncia = 4 sicilici = 8 drachmas	27.288
1 sescuncia = 1½ uncia = 6 sicilici	40.93
1 sextans = 2 unciae	54.58
i quadrans = 3 unciae	81.86
1 triens = 4 unciae	109.15
1 quincunx = 5 unciae	136.44
1 semis = 6 unciae	163.73
1 septunx = 7 unciae	191.02
1 bes = 8 unciae	218.30
1 dodrans = 9 unciae	245.59
1 dextans = 10 unciae	272.88
1 deunx = 11 unciae	300.16
1 libra = 12 unciae (also ponds or *as*)	327.45

TABLE 12. Denomination of Roman Weights and Gram Equivalents

Name	As	Ounces	Symbol	Grams
As (pondo)	1	12	I	327.45
Deunx	$\frac{11}{12}$	11	S== —	300.16
Dextans	$\frac{5}{6}$	10	S==	272.88
Dodrans	$\frac{3}{4}$	9	S== — or S== I	245.59
Bes	$\frac{2}{3}$	8	S== or —S—	218.30
Septunx	$\frac{7}{12}$	7	S—	191.02
Semis	$\frac{1}{2}$	6	S	163.73
Quincunx	$\frac{5}{12}$	5	==— or ==	136.44
Triens	$\frac{1}{3}$	4	==	109.15
Quadrans	$\frac{1}{4}$	3	==— or == I	81.86
Sextans	$\frac{1}{6}$	2	==	54.58
Sescuncia	$\frac{1}{8}$	1½	— $\mathcal{L}$ or $\mathcal{L}$—	40.93
Uncia	$\frac{1}{12}$	1	—	27.29
Semuncia	$\frac{1}{24}$	$\frac{1}{2}$	$\mathcal{L}$ (Σ) —	13.64
Sicilicus	$\frac{1}{48}$	$\frac{1}{4}$	Ɔ	6.82
Sextula	$\frac{1}{72}$	$\frac{1}{6}$	Ɩ or ∿	4.55
Dimidia sextula	1/144	$\frac{1}{12}$	Ɨ or ∿	2.27
Scripulum	1/288	$\frac{1}{24}$	Ƭ (Ɔ, ƗƗ)	1.14

TABLE 13. Roman Weights from the Classical Period*

Name	Grams	Ounces
libra (as)	327.45	12
deunx	300.16	11
dextans (decunx)	272.88	10
dodrans	245.59	9
bes	218.30	8
septunx	191.02	7
semis	163.73	6
quincunx	136.44	5
triens	109.15	4
quadrans (teruncius)	81.86	3
sextans	54.58	2
sescuncia	40.93	$1\frac{1}{2}$
uncia	27.288	1
semuncia	13.644	$\frac{1}{2}$
binae sextulae (duella)	9.096	$\frac{1}{3}$
sicilicus	6.822	$\frac{1}{4}$
sextula	4.548	$\frac{1}{6}$
drachma	3.411	$\frac{1}{8}$
dimidia sextula	2.274	$\frac{1}{12}$
scripulum	1.137	$\frac{1}{24}$
obolus	0.568	$\frac{1}{48}$
siliqua	0.189	1/144

* From Pauly, vol. 17, p. 620.

TABLE 14. Medical and Veterinarian Weight Standards of Ancient Rome According to Agricola*

Physicians: *Pondera medica* (see also the caption of Fig. 95)

Granum	Siliqua	Semiobolus	Obolus	Scripulus†	Drachme	Uncia	Libra	Mina
4	Siliqua							
6	$1\frac{1}{2}$	Semiobolus						
12	3	2	Obolus					
23	6	4	2	Scripulus†				
72	18	12	6	3	Drachme			
576	144	96	48	24	8	Uncia		
6,912	1,728	1,152	576	288	96	12	Libra	
9,216	2,304	1,536	768	384	128	16	$1\frac{1}{3}$	Mina

Veterinarians: *Pondera hippoiatrica* (mulomedicorum)

Obolus	Scripulum†	Drachme	Uncia	Libra	Mina
2	Scripulum†				
6	3	Drachme			
65	$22\frac{1}{2}$	$7\frac{1}{2}$	Uncia		
540	270	90	12	Libra	
675	$337\frac{1}{2}$	$112\frac{1}{2}$	15	$1\frac{1}{4}$	Mina

* The libra = 327.45 grams; the obolus = 10 grains.
† In classical Latin this weight is called *scripulum*, *scripulus*, or *scriptulum* (Georges, 1880, vol. 2, p. 2279).

5a. Roman Veterinarian and Medical Weights*

Under Constantine the Great the new gold coin unit, the solidus, was introduced. The weight of a solidus was ordered to be 1/72 part of a Roman pound of 327.45 = 4.548 gm. This unit soon became a very popular one in the Roman and Byzantine Empire under the name of nomisma as the standard weight unit in daily commerce.

Up to the early nineteenth century most places in Germany used these identical weights—the so-called Nuremberg apothecary weights—based on the Nuremberg silver weight standard: 2 lots (silver weight) = 1 ounce (uncia) of the medicinal or apothecary weight (Chelius, 1808, p. 91, according to personal communication from the Nuremberg Aichmeister).

6. Ancient Hindu Measures of Weight and Capacity†

Measures of Weight

The Amarakosha mentions measures of three kinds—weight, length, and capacity.

The krishnala (gunja, raktika, the black- and red-berry of the shrub *Abrus precatorius*) was employed as a natural measure of weight; 80 krishnala berries on the average weigh 105 grains Troy, and this must be taken as the basis of our computation, though in current practice 80 krishnalas are taken to be equivalent to 210 grains.

The conventional measures were, however—one gold masha was the weight of 5 krishnalas of gold, 1 suvarna or tola weighed as much as 16 mashas, and one pala as much as 4 suvarnas or tolas. A pala of gold, therefore, weighs 320 krishnalas (Manu Chap. VIII).

A masha of gold, therefore, would weigh 6 grains, a tola, 105 grains (in current practice it weighs about 180 grains); and a pala, 420 grains Troy.

1 silver masha = 2 krishnalas, 1 dharana = 16 silver mashas; 1 pala = 1 dharanas. 1 krishnala = 1296 trasarenus.

1 trasarenu, a measure of weight, therefore, is the equivalent of 7/6912 of a grain Troy, or double this according to current measures.

Raju (Part 1, 1962) lists the following weights from the Sanskrit literature (from *Manu Smriti*, Chapter 8):

---

* According to Agricola (1555, pp. 125, 132).

† Râu (1956, pp. 284–85).

Gold: 8 trasrenu   = 1 liksha

Gold:	8 trasrenu	= 1 liksha
	3 liksha	= 1 black mustard
	3 black mustard seeds	= 1 white mustard (6.89 mg)
	6 white mustard seeds	= 1 barley-corn (41.33 mg)
	3 barley-corns	= 1 krishnala (124 mg)
	5 krishnala	= 1 masha (620 mg)
	16 masha	= 1 suvarna (9.92 g)
	4 suvarna	= 1 pala or nishka (39.680 g)
	10 pala	= 1 dharana (396.8 g)
Silver:	2 krishnala	= 1 masha (248 mg)
	16 masha	= 1 dharana or purana (3.968 g)
	10 dharana	= 1 shatamana (39.680 g)

Copper: 1 karshapana $=$ 16 masha $=$ 80 krishnala $= \frac{1}{4}$ pala $=$ (9.920 g)

Weights in South India

10 seeds of masha (*Phraseolus radiarus*) or 5 seeds of gunja (*Abrus precatorius*)	= 1 suvarnamasha (620 mg)
16 suvarnamasha	= 1 suvarna or karsha (9.920 g)
4 karsha	= 1 pala (39.680 g)
88 white mustard seeds	= 1 silver masha (606 mg)
16 silver masha or 20 saibya seeds	= 1 dharana (9.696 g)
20 grains of rice	= 1 dharana of diamonds

7. Old Islamic weights*

quámha	= 0.05 gm
habba	= 0.07 gm
qirath	= 0.2 gm (0.195 gm)
dirhem	= 3.25 gm (3.125 gm = 48.225 grains)
mitqual = 1.5 dirham	= 4.875 gm (4.464 gm = 68.888 grains)
uqijje = 12 dirham	= 39 gm
ratl = 144 dirham	= 468 gm (337.55 gm)
oqa	= 1.254 kg
qantar	= 44.688 kg

These values can be regarded only as ideal standard values. Indeed the value of medieval weights (and as the next sections show, also in later periods) has been different in different places and for different commodities. Some of the names and values of Islamic weights collected from the valuable studies of Hinz have been integrated into the standards of the catalogue of weights.

---

* According to Agricola (1555) and quoted by Alberti (1957, p. 32). The standard figures, as determined by Hinz (1956, pp. 2 ff.) are added in parentheses.

APPENDIX 2

Weight Standards of the World before the Metric-Decimal System

Before listing the early weight standards we define a few of the most important weighing standards of Europe and their subdivisions, used before the metric-decimal era.

1. The Mark of Cologne

This standard unit is first mentioned as such in a contract between Frederic Barbarossa and the Count of Barcelona and Provence in 1162: "Quindecim marcas auri boni ad justum pondus Coloniense" (Fifteen marks of good gold according to the just weight [unit] of Cologne). (See *Monumenta Germaniae*, vol. 1, p. 305; Art. 4; also Guilhiermoz, 1919, p. 21.) It was the standard unit of Germany and of most countries in Europe up to the middle of the nineteenth century.

According to Chelius (1805, pp. 68, 142) [1 pfund (pound) = 2 marc = 467.62 gm].

$$1 \text{ marc} = 16 \text{ loth} = 152 \text{ engels} = 65{,}536 \text{ richtpfennig} = 67 \text{ ducats}$$
$$1 \text{ loth} = 4 \text{ quint} = 4{,}096 \quad \text{,,}$$
$$1 \text{ quint} = 4 \text{ pfennig} = 1{,}024 \quad \text{,,}$$
$$1 \text{ pfennig} = 256 \quad \text{,,}$$
$$16 \text{ pfennig} = 18 \text{ grains}$$

According to a protocol signed by the sealer (*Gewichtsaicher*) and assayer of Frankfurt a/M, signed on September 30, 1771, the mark of Cologne was also equal to 4,020 asse; 1 ducat = 60 asse, and one ducat-ass = 16.30 richtpfennig (Chelius, 1808, p. 74).

2. The French Pound (livre)

According to Machabey (1959, p. 359), and Chelius (p. 68)

livre	1						489.506 gm = 5,094 Dutch asse
marc	2	1					244.753 gm
onces	16	8	1				30.594 gm
gros	128	64	8	1			3.823 gm
deniers	384	192	24	3	1		1.274 gm
grains	9,216	4,608	576	72	24	1	0.0531 gm

3. Weights of Great Britain

These are from English conversion tables (no year, no author) printed in the eighteenth century (found in the Medical Historical Library of Yale University).

Troy Weights

24 grains = 1 penny weight
20 penny weights = 1 ounce
12 ounces = 1 pound [= 373.246 gm]
14 ounces 11 penny weights 20 grains = 1 lb. averdupoise

Averdupoise Weights

$27\frac{1}{3}$ grains = 1 dram
16 drams = 1 ounce
16 ounces = 1 pound [= 453.59 gm]
$13\frac{1}{8}$ ounces = 1 pound Troy
14 pounds = 1 stone

Trone or Old Scotch Weights

20 ounces = 1 pound
16 pounds = 1 stone

Listed below are cities and countries of the world and the names of their standard units of weights used before the introduction of the metric-decimal system. These data are selected mainly from Niemann (1830), Nelkenbrecher (1848), Noback (1877,) Oehlschläger (1857), and Aubök (1893). If only the name of the unit is given, see details in Appendix 3.*

Aachen (Germany)   1 pound = 32 loth = 1/100 centner = 467.04 gm.
Abyssinia   1 rottolo = 12 wakeas = 120 drachms = 311 gm.
Achem (Sumatra)   1 bähar (kandil) = 200 kettih (see Catti); 1 catti = 20 bunkals = 100 taehls = 200 pagodas = 1,600 meh = 960.2 gm; 1 pagoda = 8 meh = 32 copang.
Aleppo (Syria)   1 rottolo (rotto) = 12 ounces = 2.217 kg = 720 drams; the rottolo for Syrian silk = 700, for Persian silk 680 drams; each dram = 3.5 gm; 1 oka = 400 drams.
Alexandria   1 cantaro = 100 Egypt rotoli = 44.6 kg; 1 rotolo (pound) = 12 ounces = 144 oka-drams; gold and silver weight: 1 oka-dram = 16 kirat = 64 grains.

---

* The various authors quoted occasionally give different gram and kilogram values for these weights, but the differences are as a rule not very large.

Algiers   1 unze = 8 derahem = 34.13 gm = 1/16 rotel a'thary;
                       for silver = 1 rotel fedhy = 497.435 gm; for gold
                       = 1 metskal = 4.67 gm; for diamonds = 1 gyrath
                       (carat) = 0.207 gm.
Alicante (Spain)   1 big pound = 517.29 gm; 1 small pound = 344.89gm;
                       1 quintal = 4 arrotas = 96 big pounds of 18 ounces
                       or 36 small pounds of 12 ounces.
Altona   See Hamburg.
Amsterdam   1 pond = 1 kg = 10 oncen = 100 looden = 1,000
                       wigtjes = 10,000 korrels; 1 medicinal pound = 375
                       gm (wigtjes) = 12 ounces = 96 drachms = 288 scrupels
                       = 5,760 grains; old Dutch weight: 1 troy pound =
                       492.168 gm = 2 mark = 16 ounces = 320 engelsen =
                       10,240 *as*.
Antwerp   Commercial weight: 1 charge = 2 bales = 400 pounds;
                       1 ship's pound = 300 pounds; 1 centner = 100 pounds;
                       1 chariot = 165 pounds; 1 stone = 8 pounds; 1 pound
                       = 2 mark = 16 unzen = 32 loth; 1 loth = 10 engel;
                       1 pound = 468.8 gm; 1 quintal = 100 pounds.
Aragon   1 pound = 349.8 gm; 1 quintal = 4 arrobas = 144
                       pounds; gold and silver weight: 1 mark = 230 gm.
Augsburg   1 pound (lightweight) = 472.4 gm; 1 pound (heavy-
                       weight) = 490.87 gm; silver weight: 1 pound = 2 marc;
                       1 marc = 16 loth = 64 quint (quentchen) = 256
                       pfennig; 1 mark = 235.9 gm.
Austria   See Vienna.
Balua or San   See Rio de Janeiro.
  Salvador
Bangkok   See Sumatra.
Basel   1 centner = 100 pounds; 1 pound = 32 loth; 1 heavy
                       pound = 493.24 gm (commercial weight); 1 pound
                       (retail) = 486.2 gm; silver weight: 1 pound = 467.71
                       gm; gold weight: 1 crown = 3.37 gm.
Batavia   1 pecul = 100 catties = 1,600 taels = 49.052 kg; 3
                       pecul = 1 small bahar, $4\frac{1}{2}$ pecul = 1 large bahar.
Bavaria   1 pound = 32 lot = 128 quentchen = 560.06 gm.
Belgium   See Antwerp.
Bengal   Bazaar weight: 1 maund = 40 seer = 640 chittaks =
                       3,200 siccas; for different goods 1 maund is between
                       30.6 and 37.2 kg; for gold and silver: 1 sicca = 10
                       massa = 80 ruttees = 15 gm.

Berlin	1 Prussian pound = 467.7 gm; commercial weight: 1 centner = 110 pounds = 3,520 loth = 14,080 quentchen; 1 customs centner (Zollcentner) = 100 customs pounds (Zollpfund) = 50 kg; 1 shipping ton (Schiffslast) = 4,000 pounds.
Bohemia	See Prague.
Bologna	1 pound (libbra) = 361.85 gm; 1 peso = 25 pounds; 1 pound = 12 ounces, 1 ounce = 16 ferlini.
Bombay	1 candy = 20 (maon) maunds; 1 maund = 33.86 or 37.25 kg = 40 seers; 1 seer = 30 pice = 72 tauk; gold and silver weight: 1 tola = 40 vall; pearl weight: 1 tauk = 24 ruttees = 96 quarter = 384 annas = 4.67 gm.
Boston, Mass.	See London.
Bremen	1 pound = 498.5 gm; 1 centner = 116 pounds commercial weight; 1 pound = 32 loth = 128 quentchen = 512 orth; the freight pound = 300 commercial pounds; the heavy pound or quintal of land freight = 300 pounds or 22 liespfund at 14 pounds = 308 pounds; 1 retail pound (Krämergewicht) = 470.28 gm.
Brussels	1 commercial pound = 467.67 gm; 1 pound mark weight (Markgewicht) = 492.15 gm.
Bucharest	1 oka = 1,262.9 gm; 1 cantar = 44 oka.
Buenos Aires	1 libra = 2 marcos = 16 onzas; 1 onza = 16 adarmes = 567 granos; 1 adarme = 1.75 gm; 1 libra = 9.216 granos; 1 quintal = 4 arrobas = 100 libras.
Bukhara	1 batman = 127.767 kg.
Calcutta	1 maund = 40 seers; 1 seer = 16 chattacks = 80 siccas or tolahs; 1 factory-maund = 33.86 kg; 1 bazaar-maund = 37.25 kg; gold and silver weight: 1 sicca = 10 massa; 1 massa = 32 grän; 1 gran = 4 punkhos = 11.64 gm; 25 grän = 1 anna; 16 annas = 1 tolah = 14.55 gm; grain weights: 1 khahoon = 40 maunds; 1 maund = 16 soalis = 320 pallies; 1 pallie = 4 raiks = 16 koonkes = 80 chattacks = 4,119 gm.
Canada	See London.
Canton (China)	1 tael or tale or lyang = 37.78 gm; 1 pecul or picol = 100 catties = 60.478 kg; 1 catty = 16 taels; 1 tael = 10 maces; 1 mace = 10 candarines; 1 candarine = 10 cashes (the catty is also called gin; the tael, lyang; the mace tachen or tsien; the candarine or condorine

227

	is also twen, swin, or fuen; the cash, li); gold and silver weight: 1 catty or gin = 16 taels; 1 tael = 10 cheh = 100 hoon = 1,000 li = 10,000 si = 100,000 hoot.
Chile	1 libra = 2 marcas = 16 oncas = 1 English avdp. pound; 1 gold marca = 4,800 Spanish granos; 1 silver marca = 4,608 granos (grains).
China	See Canton. See values, denominations and symbols in Chart 10.
Cologne	1 pound = 467.62 gm; 1 centner = 106 pounds; 1 pound = 32 loth; the true Cologne mark = 233.81 gm; 1 mark = 16 loth = 64 quint = 152 engels = 256 pfennig = 65,536 richtpfennigtheilche.
Constantinople	1 oka = 1,278.5 gm = 4 cheky = 400 drachmas; 1 teffé silk = 600 drachmas; 1 batman = 6 oka; 1 cantaro = 44, also 45 oka = 100 rotoli.
Copenhagen	1 pound commercial weight = 500 gm = 16 onces = 32 loth = 128 quentchen (quintin) = 512 ort = 8,192 es.
Damascus	1 kantar = 100 rotoli; 1 rotolo = 60 ounces = 1,785 gm; gold and silver weight: 1 ounce = 29.75 gm.
East India	See Batavia and Bengal.
Egypt	See Alexandria.
England	See London.
Fiume	1 pound = 558.76 gm.
Florence	1 libbra = 339.5 gm = 1/100 centinajo = 12 onci = 288 denari = 6,912 grain; 1 libbra of 16 ounces = 453.9 gm.
France	Up to end of the 18th century the French pound (livre) of Paris = 16 ounces and was subdivided in two ways: (i) 1 livre commune = 489.5 gm = 2 marc = 16 ounces = 128 gros = 384 deniers, each denier = 24 grains; (ii) 1 livre = 2 demilivres = 4 quarterons = 8 demiquarterons = 16 ounces = 32 demiounces. The first was used for valuable goods. Lyons: 100 livres = 88 Paris livres; the silk pound of Lyons = 15 ounces only. In Toulouse and Haut Languedoc the pound (poid de table) = 13.5 ounces of the Paris pound and in Marseille and Provence 1 pound = 13 ounces of the Paris pound. For the metric-decimal system, see Paris.

Frankfurt-am-Main 1 old pound = 467.91 gm; 1 new pound = 467.711 gm; 1 light pound = 2 mark = 32 loth = 128 quentchen = 512 pfennig; 1 heavy pound = 505.35 gm; ducat weight: 1 marc = 67 ducats = 4,020 ducat-asse.

Geneva (i) The heavy pound (gros poids) = 18 onces = 550.69 gm; 1 once = 24 deniers. (ii) The light pound (petit poids) for silk = 458.91 gm = 15 ounces. (iii) The marc weight (poids de marc): 1 pound = 489.506 gm = 16 ounces; gold and silver weight: 1 marc = 8 ounces = 64 gros = 192 deniers at 24 grains each = 244.753 gm; apothecaries weights: 1 pound (livre) = 16 ounces = 128 drams = 384 scruples = 9,216 grains = 500 gm.

Genoa 1 heavy pound (peso grosso) = 348.687 gm; 1 light pound (peso sottile) = 316.97 gm; 1 rubbio = 25 libbre = 7.9195 kg; 1 cantaro = 47.517 kg; 1 peso = 5 cantari = 30 rubbi = 500 rotoli = 750 libbre = 900 oncie. According to Aubök, 1 libbra = 368.845 gm.

Great Britain See London.

Guinea (West Africa) 1 benda = 8 pisos or usanos (unzen) = 989.5 English grän = 64.11 gm.

Haiti See France.

Hamburg 1 Hamburg pound = 484.6 gm = 32 loth = 128 quentchen = 512 pennyweight; 1 ship's pound = $2\frac{1}{2}$ centner = 20 lies pounds = 280 pounds; 1 stone flax = 20 pounds; 1 stone of wool or feathers = 10 pounds; 1 centner = 112 pounds; the Hamburg ship's last = 4,000 pounds.

Holland See Amsterdam.

India 1 mound = 40 seers = 37.3 kg; 1 tola = 12 masha = 96 reti = 11.7 gm.

Japan 1 picul or pecul = 100 cattis = 57.962 kg; The standard unit according to imperial decree of March 23, 1891, was 1 kwan = 3,750 gm; 1 mé or monmé = 1/1,000 kwan = 10 fun = 100 rin = 1,000 mô; 1 kin (pound) = 600 gm; 1 monmé = 3.75 gm; 1 fun = 0.375 gm; 1 rin = 0.0375 gm; 1 mô = 0.00375 gm; apothecary weights: 1 rjoo = 4.3 monmé; 1 maj = 10 rjoo = 43 monmé.

Java See Batavia.

Kraków (Poland) 1 stein = 32 pounds; 1 pound = 404.93 gm.

Leghorn See Florence.

Leipzig — 1 centner = 5 stone (stein) = 110 pounds; 1 pound = 32 loth = 128 quintchen = 512 pennyweight = 1,024 heller-weight = 467.54 gm; 1 ship's pound = 3 cwt (centners).

Lemberg (Poland) — 1 pound = 420 gm; 1 centner = 100 pounds.

Lisbon — 1 quintal = 4 arrobas = 1 arroba = 32 arratels or libras (pounds); 1 libra = 2 marcas; 1 pound (libra or arratel) = 459.1 gm; 1 arratel = 2 meios = 4 quartas = 16 oncas = 128 oitavas = 384 scruples; 1 scruple = 24 grains; the gold and silver marco = 8 oncas = 64 oitavas = 192 escrupulos = 4,608 granos = 229.55 gm; as jewel weight 1 quilat or karat = 4 granos = 0.2058 gm; apothecary weights: 1 arratel = 1½ marcos of the gold weight or ¾ pound of the commercial weight = 12 oncas = 96 oitavas = 288 escrupulos = 6,912 granos.

London — Troy weight serves as gold, silver, mint, diamond, and apothecary weight; avdp. pound as commercial weight: 1 troy pound = 12 ounces = 373.246 gm; 1 avdp. pound = 16 ounces = 453.59 gm; 1 troy pound = 12 ounces = 240 pennyweight = 5.760 grains; as apothecary weight: 1 pound = 12 ounces = 96 drams = 288 scruples = 5.760 grains; 175 troy pounds = 144 avdp. pounds, but 175 troy ounces = 192 avdp. ounces.

Lucca — 1 libbra (della grascia) = 334.5 gm; 1 libbra (della commissione) = 341 gm.

Lucern — 1 pound = 36 loth; 1 loth = 4 quintlein = 528.9 gm; 1 medical pound = 357.95 gm.

Lübeck — 1 commercial pound = 486.47 gm; 1 pound = 32 loth; 1 loth = 4 quent; 1 centner = 112 pounds = 8 lies pounds; 1 ship's pound = 20 lies pounds; 1 lies pound = 14 pounds.

Lyons — 1 poids de marc = 489.8 gm; 1 poids de soie = 458.912 gm; 1 poids de ville = 418.76 gm.

Madras — Commercial weight; 1 candi = 20 maunds = 226.78 kg; 1 maund = 11.33 kg = 8 vis; 1 vis = 5 seer; 1 seer = 8 paloins; 1 paloin = 10 pagodas; 1 pecul = 132 pounds = 59.875 kg; gold and silver weight: English troy pound is used; also, star-pagoda = 3.405 gm.

Madrid — 1 Castilian mark = 230.07 gm; 1 mark = 8 oncas = 64 ochavas = 128 adarme = 384 taminos = 4,608 granos; 1 quintal macho = 6 arrobas = 150 libras; 1 libra = 2 marcos = 4 quarterones = 16 oncas.

Majorca	Commercial weight: 1 pound = 12 ounces = 408 gm; 1 quintal = 4 arrobas at 25 pounds.
Makassar	See Batavia.
Malta	1 rotolo = 791.59 gm; 1 cantaro = 100 rotolo; 1 libbra or pound as gold and silver weight = 316.61 gm.
Marseille	1 livre poids de table = 16 onces = 407.93 gm; 1 once = 8 gros; 1 gros = 72 grains; 100 livres = 1 quintal; 3 quintals = 1 last.
Mexico	See Spain.
Mocha (Yemen)	1 wakeia (wakega) = 31.6 gm; 1 maund = 1.241 kg; 1 mokka pound = 404.67 gm.
Modena	1 quintale = 100 libbre; 1 libbra = 12 once; 1 once = 16 ferlini; 1 libbra = 340.457 gm.
Montpellier (France)	1 ship's last = 2,000 kg. See also Nîmes.
Morocco	Commercial weight: 1 pound = 20 Spanish piasters; 100 pounds = 1 quintal = 53.973 kg; 1 rottolo = 526.3 gm.
Munich	1 pound = 560 gm; 1 centner = 5 stein = 100 pounds = 3,200 loth = 12,800 quentchen.
Naples	1 libbra = 12 once = 320.759 gm; 1 rottolo = 33.3 once; 1 cantaro grosso = 100 rottoli = 89 kg; 1 cantaro piccolo = 100 libbre; gold, silver and silk weight: 1 gold oncia = 24 carati; 1 carat = 100 parti; 1 silver libbra = 12 denari; 1 denaro = 100 parti.
Neufchatel (Switzerland)	Poids de fer or commercial weight: 1 pound = 520.1 gm = 2 marc; 1 marc = 8 onces; 1 once = 8 gros; 1 gros = 3 deniers; 1 denier = 24 grains; 1 quintal = 100 pounds.
New York	See London.
Nîmes and Montpellier	Livre grosse = 411.1 gm; 100 livre grosse = 1 cantar; libre subtile = 318.24 gm.
Norway (also Sweden and Iceland)	1 øre = 26.805 gm. = 7 drachmer (each 3.828 gm) = 28 scripula (each 0.957 gm).
Nuremberg	One pound commercial weight = 509.996 gm; 1 pound silver weight = 477.138 gm; 1 Nuremberg mark = $\frac{1}{2}$ pound silver weight = 238.569 gm; 1 medicinal pound

	= 357.854 gm = 12 unzen = 96 drams = 288 scruples = 5.760 gran.
Padua	1 peso grosso = 486.539 gm; 1 peso sottile = 338.88 gm.
Palermo	Commercial weight: 1 cantaro grosso = 100 rotoli grossi; 1 rotolo grosso = 33 once; 1 cantaro grosso = 110 rotoli sottili; 1 rotolo sottile = 30 once; 1 cantaro grosso = 275 libbre; 1 libbra = 12 once; 1 cantaro sottile = 100 rotoli sottili; 1 rotolo sottile = 30 once; 1 cantaro sottile = 250 libbre; 1 libbra = 12 once; 1 rotolo of 33 once = 873.268 gm; 1 rotolo of 30 once = 793.88 gm; 1 libbra of 12 once = 360 trappesi = 317.552 gm; gold and silver weight: 1 libbra (317.55 gm) = 2/5 rotoli = 12 once = 96 dramme = 288 scrupoli = 5,760 grani = 46,080 ottavi.
Paris	Metric system since 1840 obligatory; 1 myriagramme = 10 kilogrammes = 100 hectogrammes = 1,000 décagrammes = 10,000 grammes; 1 gramme = 10 décigrammes = 100 centigrammes = 1,000 milligrammes; 1 quintal = 100 kg; 1 millier = 1,000 kg. In the transition time before the end of 1839: 1 livre usuelle = 500 gm = $\frac{1}{2}$ kg = 4 quarterons = 16 onces = 128 gros. The old weight system: 1 pound (poids de marc) = 489.506 gm; 1 livre = 2 marc = 16 onces = 128 gros = 9.216 grains; 1 denier = 24 grains; 1 quintal = 100 pounds; 1 millier = 1,000 pounds, or 10 quintaux = 489.5 kg; 1 charge = 3 quintaux; 1 ship's last = 2 milliers = 20 quintaux = 2,000 livres; gold weight and silver weight: 1 marc = 8 onces; 1 once (30.59 gm) = 8 gros = 24 deniers = 576 grains; apothecary weight: 1 livre = 16 ounces = 128 drachmes = 384 scruples = 9,216 grains.
Parma	1 pound (libbra) = 328 gm = 12 once = 288 denari = 6,912 grani; 1 rubbo = 25 libbre.
Persia	1 Tauris mound = 6 rattal = 300 derhem = 1,200 miskal = 6.34 pounds avdp.; 1 mismal = 4.59 gm (Aubök).
Pesth (Hungary)	See Vienna.
Prague	See Vienna. The old Bohemian weight: 1 centner = 6 stein; 1 stein = 20 pounds; 1 pound = 514.35 gm.
Prince of Wales Island (Malaya)	Commercial weight is the Chinese catty = 16 tales = 604.8 gm; 100 catties = 1 centner or pecul; 3 peculs =

232

	1 bahar; 4,000 catties or 40 peculs = 1 coyan; gold and silver weight: 1 buncal = 16 miams = 45.79 gm.
Prussia	See Berlin.
Ragusa (Yugoslavia)	See Vienna.
Regensburg	See Munich.
Reval (Estonia)	See St. Petersburg.
Riga	Has used Russian weights since 1845; older weights: 1 ship's pound = 20 lies pounds; 1 lies pound = 20 pounds; 1 pound = 32 loth; 1 loth = 4 quent; 1 pound = 418.83 gm.
Rio de Janeiro	Commercial weight: 1 quintal = 4 arrobas; 1 arroba = 32 pounds (libras); 1 pound = 2 marcos; 1 marco = 8 oncas; 1 onca = 8 oitavas; 1 oitava = 3 escrupulos; 1 escrupulo = 24 grãos; 100 pounds = 45.875 kg; 1 arroba (a) = 14.685 kg; 1 quintal (ql) = 58.741 kg.
Rome (Vatican)	Commercial weight: cantaros of 100, 160, and 250 pounds; 1 cantaro grosso = 10 cantari piccoli = 100 decima; 1 centenaio = 100 pounds; 1 migliajo = 1,000 pounds; also used as gold, silver, coin, and medicinal weight: 1 libbra = 12 once = 288 denari = 6,912 grani = 339.156 gm; as medicinal weight: 1 oncia = 8 dramme = 24 scrupoli = 576 grani.
Rostock (Germany)	1 Mecklenburger pound = 484.726 gm; 1 Rostock pound (Stadtwaagegwicht) = 508.229 gm; 1 Rostock pound (Krämergewicht) = 484.028 gm; 1 centner = 112 pounds; 1 pound = 32 loth; 1 loth = 4 quentchen; 1 ship's pound = 2½ centner = 20 lies pounds; 1 lies pound (=14 pounds) in Rostock = 16 pounds; 1 last = 2 tons; 1 ton = 20 centner; 1 centner = 100 pounds; 1 ship's pound = 20 lies pounds; 1 lies pound = 16 pounds; 1 stein of wool or feathers = 10 pounds; 1 stein flax = 20 pounds.
Rotterdam	See Amsterdam.
Russia and Ruthenia	See St. Petersburg.
St. Gallen (Switzerland)	1 centner = 100 pounds; 1 pound heavyweight = 40 loth = 577.702 gm; 1 pound lightweight = 32 loth = 465.127 gm.
St. Petersburg	1 pound = 409.517 gm = 12 lana = 32 loth = 96 solotnik = 9,216 (parts) doli; 1 pud = 40 pounds; 1

	pud = 4 desacterick = 8 parterick = 13½ troinik = 20 dwoinik. According to Leutmann (1729), the Ruthenian weights are: 1 percoiwitz (berkowetz) = 10 pud = 400 pounds; the medicinal pound = 358.328 gm.
Santiago de Chile	See Chile.
Saragossa (Spain)	1 quintal = 4 arrobas; 1 arroba = 36 libras; 1 libra = 345.10 gm.
Sardinia	See Genoa, Turin.
Siam	See Bangkok.
Sicily	See Palermo.
Singapore	Commercial weights are the Chinese catty, the pecul, coyan, etc.; 1 pecul = 60.47 kg; gold and silver weight: 1 buncal = 2 Spanish silver piasters.
Spain	See Alicante, Aragon, Madrid, Saragossa, Valencia.
Stockholm	1 skeppund = 20 lispund = 400 schalpfund, skalpund, or marc; 1 pound = 425.34 gm; 1 pound iron weight = 340.2 gm; 1 pound mining weight = 375.9 gm; 1 pound raw iron = 488.7 gm; 1 pound raw copper = 377.5 gm; 1 mark gold and silver weight = 210.64 gm = 32 loth; 1 loth = 4 quentin; 1 quentin = 68½ Swedish *as*; 1 medicinal pound = 356.437 gm.
Stuttgart	1 pound = 467.728 gm; 1 centner = 104 light pounds. = 3,328 loth = 13,312 quentchen = 100 heavy pounds.
Sumatra	See also Achem. Commercial weight: 1 bohar = 220 Malayan or 330 Chinese catties = 202.99 kg; 1 Malayan catty = 922.7 gm; 1 Chinese catty = 615.13 gm; gold and silver weight: 1 tale = 16 maces = 853.3 Dutch *as*. On the Isle of Banka the commercial weight is: 1 cojäng = 80 balys; 1 baly = 20 gantangs; 1 gantang = 6 katjes; 1 cojäng = 6,000 pounds Dutch troy = 2,952.6 kg; as gold and silver weight: 1 catty = 10 tales; 1 tale = 2¼ Batavian reals = 1,280 Dutch *as* = 61.52 gm.
Surat (East India)	1 candy = 20 maunds; 1 maund = 40 seers; 1 seer = 30 pice; 1 maund = 16.933 kg; gold and silver weight: 1 tola = 32 valls; 1 vall = 3 ruttees = 784 Dutch *as* = 03.768 gm.
Surinam	See Amsterdam (old Dutch weights).
Switzerland	See Basel, Geneva, St. Gallen.
Sydney	See London.
Syria	See Aleppo, Damascus.

Teheran	See Persia.
Tiflis	See St. Petersburg.
Trier	1 commercial pound = 467.69 gm; 1 medicinal pound = 356.12 gm.
Trieste	Commercial weight used for purchase of goods was the same as the Venetian; goods sent to Germany were weighed by Viennese weights.
Tripoli	Commercial weight: 1 cantaro or centner = 100 rottoli or pounds; 1 rottolo or pound = 16 ounces; 1 ounce = 8 termini; 1 rottolo = 508.63 gm.
Tunis	Commercial weight: 1 cantaro = 100 rottoli; 1 rottolo = 16 ounces = 503.6 gm.
Turin	1 libbra (pound) = 368.845 gm = 12 once; 1 once = 8 ottavi; 1 ottavo = 3 denari; 1 denaro = 24 grani; gold and silver weight: 1 marco = 8 once; 1 once = 24 denari; 1 denaro = 24 grani; 1 grano = 24 granottini; 1 marco = 245.896 gm; medicinal weight: 1 pound = 12 once = 96 dram = 288 scrupoli = 5,760 grani = 307.37 gm.
Turkey	See Constantinople.
United States	See London.
Valencia	Commercial weights: 1 carga or carica = 3 quintals; 1 quintal = 4 arrabos; 1 arroba = 24 big pounds = 36 small pounds; coffee, sugar, spices, tobacco, and similar goods weighed by the small pound (libra sutil) = 12 onças = 48 quartos = 192 adarmes = 6.912 granos = 5,494 English grain = 356.01 gm; 1 big pound (libra gruesa) = 18 onças; gold and silver weight: 1 marca = 8 onças = 32 quartos = 128 adarmes = 4,608 granos = 3,557.6 English grain = 230.5 gm.
Venice	Old weights. Commercial weight: peso grosso (heavy pound) and peso sottile (light pound). Peso grosso: 1 pound (libbra) = 2 marks (gold and silver weight) = 12 once = 72 sazi = 2,304 carati = 9,216 grani = 476.999 gm; 1 peso sottile = 12 once = 72 sazi = 1,728 carati = 301.23 gm; 12 pounds peso grosso = 19 pounds peso sottile; 1 centinajo = 100 libbre grosse; gold, silver, and jewel weight: 1 marco = 8 once = 32 quarti = 192 denari = 1,152 carati = 4,608 grani.

235

Vienna	1 Viennese pound commercial weight (Handelspfund) = 560.012 gm = 1 handelspfund = 4 vierding = 16 unzen = 32 loth = 128 quintal; 1 centner = 5 stein = 100 pounds; 1 stein = 20 pounds; 1 saum = 275 pounds 1 lägel steel = 125 pounds; 1 saum steel = 2 lägel; 1 karch = 400 pounds; gold and silver weight: 1 Viennese mark = 280.644 gm = 16 loth = 80.4 ducats; 1 loth = 4 quint; 1 quint = 4 pfennigs; 1 pfennig = 256 richt-pfennigtheilchen; 1 Viennese medicinal pound = 24 loth commercial weight = 420.009 gm.
Warsaw	1 funt (pound) = 405.504 gm = 16 uncyi (ounces) = 32 lutow (loth) = 128 drachma; 1 centnar (centner) = 4 stein = 100 pounds; 1 medicinal pound = 358.510 gm.
Würtemberg	See Stuttgart.
Yemen	See Mocha.

APPENDIX 3

This catalogue lists the names of weights used in different parts of the world up to the middle of the nineteenth century, with their equivalents in gram or kilogram and with modern values according to the information given by Niemann (1830), Nelkenbrecher (1848), Oehlschlager (1857), Noback (1877), Aubök (1893), and others. The excellent monograph of Hinz (1955) was a source for Islamic (mainly medieval) weights; the reader is referred to this book for additional details. Articles in the Indian journal *Metric Measures* (since 1958) have been of highest value for Indian weights.

Abas	Weight unit for pearls in Persia = 0.175 gm.
Abucco, Abuhi	Gold and silver weight in Pegu, Indochina, = 196.44 gm.
Acino, Ass	Gold and silver weight in both Sicilies = 0.038 gm = 600 acini = 1 Neapolitan ounce.
Adarme	Gold and silver weight in Spain (especially Valencia): 16 adarme = 1 ounce, 128 ounces = 1 Spanish marc; 1 adarme = 36 granos = 1.75 gm; in Barcelona 1 adarme (arienzo) = 3.17 gm.
Adila	Islamic = $\frac{1}{2}$ himl = 125–150 kg.
Adowlie	Grain weight, Bombay; 16 adowlies = 1 parah, 128 (or 8 parah) = 1 candy; 1 adowlie = 1.982 kg; 1 heavy adowlie = 2.031 kg: 20 adowlies = 1 heavy parah, 125 = 1 heavy candy.
Agito, Giro	Gold and silver weight in Pegu (Indochina) = 392.9 gm = 2 abbucci = 25 tical = 1,280 moyon = 5,000 toque.
Akey	Gold and silver weight in Sudan = 1.3 gm.
Almane, Almene	Spice (saffron) weight in East India = 1 roik = 1,126.67 gm.
Amat	Batavia = 2 peculs or 200 catties or 123.5 kg.
Anna, Ana	See Bombay.

---

* If equivalents are available. References are to cities or countries listed in Appendix 2.

Aratel, Pfund	Gold and silver weight in Brazil = 2 marc or 16 ounces or 128 octavas or 384 scruples or 9,216 grains = 454.25 gm.
Argienso, Argienco	See Adarme.
Arroba	Weight or liquid unit in Spain and Portugal. Castilian arroba (in Madrid, Malaga, etc.) = 25 libras (pounds) = 50 marcos = 11.522 kg; 4 arroba = 1 cent (hundredweight) or quintal; 6 arroba = 1 quintal macho (heavy cent). Alicante and Valencia: 1 arroba = 24 libras machores (heavy pounds) = 36 libras menores (light pounds) = 432 ounces; 4 arroba = 1 quintal; 10 arroba = 1 cargo; 1 arroba = 12.452 kg. Oil arroba = 36 light pounds = 12.853 kg; 35 oil arroba = 1 pipe oil. Portugal (especially Lisbon): 1 arroba = 32 libras = 64 marcas; 4 arroba = 1 quintal; 1 arroba = 14.696 kg.
Aruzza	A grain of rice = 0.019 gm.
Ass, Esschen	Fraction of marc or pound. Dutch troy marc = 5,120 ass; 1 ass = 0.048 gm. Cologne silver marc = 4,864 ass; 76 ass = 1 quentchen, 304 = 1 loth, 608 = 1 ounce, 4,864 = 1 mark. Leipzig: gold and silver mark = 4,422 ducat-ass, each 0.056 gm. Bavaria: the silver marc = 4,352 ass, each 0.052 gm. In some places ass was also used for 1 pharmaceutical pound.
Bahar, Bähar, Bazer, Bar, Bhar, Baar, Baer	East India. Amboina. A behar of cloves = 270.658 gm. Batavia; small bahar = 3 pecul = 300 catties = 185.223 kg; large bahar = 4½ pecul = 450 catties = 277.834 kg. Prince of Wales Island: = 181.420 kg. Sumatra and Achem, 1 bahar = 200 catties = 400 buncal = 20,000 taels = 40,000 pagodas = 192.026 kg. Also an Islamic weight changing value according to commodity from 207.9 to 420.9 kg (see Hinz, pp. 91).
Baia	Islamic = 4.5 kg.
Bale, Ballen	Market weight in Brussels, Antwerp, and Netherlands = ½ charge = 2 cent (hundred pounds) = 1,950,800 Dutch ass = 93.732 kg.
Baly	See Sumatra.
Baquila	Egyptian (bean) = 4 samuna = 12 qirat = 2·34 gm.
Barbaresco, Cantaro Barbaresco	Hundred-pound weight of Mallorca; see cantaro.

238

Batmar, Batman, Männ	Two types of weights used in Turkey and Persia. (i) Big batman: $7\frac{1}{3}$ = 1 cantaro or cent; this batman = 4 small batman = 8 small oka = 16 rottel = 32 cheky = 10.205 kg. (ii) Small batman: $29\frac{1}{3}$ = 1 cantaro; this batman = 20 kas = 4 rottel = 8 cheky = 2.551 kg. (iii) Constantinople: batman to weigh silk = 6 oka = 24 cheky = 3.827 kg. (iv) Persia: 100 batman = 1 kalvar and the batman of Miranda = 5,942 gm = 640 derhem = 1,280 miskals = 7,680 dungs. Big batman of Tauris = 500 derhem = 1,000 miskals = 6,000 dungs = 4,642 gm. Small batman of Tauris = 470 derhem = 940 miskals = 5,640 dungs = 4,363 gm.
Bazaar maund	See Calcutta.
Benda	Central Africa = 1,334 Dutch ass = 64 gm.
Berkowitz, Berkowetz, Bercherect	Russian ship pound = 10 pud = 400 pounds = 384,000 solotniks = 163.597 kg.
Bis	Market weight in Pegu (Indochina) = 1,536.9 gm. Coromandel: 1 bis = 1,369.54 gm.
Bismar pound, Bissmer pound	Copenhagen = 12 Danish pounds = 5,993 gm; 3 bismar pounds = 1 wog.
Blanc	See Moneyers' weight.
Bogca	Islamic = 4 batman = 1,580 Osmanic dirham = 20.268 kg.
Bohar	See Sumatra.
Buncal, Bunkal	Gold and silver weight in Indochina: Singapore and Prince of Wales Island, 45.38 gm; 1 buncal = 16 mians; Sumatra, 20 buncals = 1 catti, 4,000 buncals = 1 bahar; 1 buncal = 48 gm = 5 tal = 10 pagodas = 320 copang.
Buttima	Persian = 30 ratel = 11.482 kg.
Cafla	Gold and silver weight in Arabia (Mocca) = 16 crat; 10 cafla = 1 wakega; 1 cafla = 3.167 gm.
Candarine, Condorin	See Canton. Swin, Fuen.
Candil, Candy	Market weight in East India = 20 maons = 160 vis = 800 seers = 2,000 paloins = 24,000 pices = 254.029 kg. Weight for grain or rice: 1 candy = 8 parahs = 128 adowlies = 512 seers = 1,024 tiprees = 210.636 kg. Calcutta: 1 candy = 218.563 kg. Madras: 1 candy = 226.806 kg. Surat: 1 candy = 338.283 kg. For wool, 1 candy = 21 maons = 840 seers = 352.7 kg.

239

Cantar, Quintal,   See Syria.
  Cent, Cantaro

Cantaro, Cantarelli,   Sardinia = 100 libra = 1,200 ounces = 40.118 kg.
  Quintal   Cagliari = 104 libra = 1,248 ounces = 41.034 kg. In Turkey and Italy commercial weight. In Spain, a measure of wine. In Syria two types of cantari: one = 100 rottoli, the other 175 rottoli. Copper, balm, camphor, and other valuable substances weighed with the rottolo of Damascus = 600 drachmas; this cantaro = 189.988 kg. Persian silk cantaro = 215.319 kg. Syrian silk cantaro = 221.653 kg. Cotton and less valuable merchandise cantaro = 227.987 kg. Cantaro of Tripoli = 175 rottoli; 1 rottolo = 720 drams; this cantaro = 398.872 kg. Alexandria and Cairo: common cantaro = 100 rottoli = 42.394 kg. Algiers: copper and wax cantaro = 54.050 kg; cotton and almond cantaro = 59.455 kg; for iron, lead, wool, 1 cantaro = 81.075 kg; for oil, soap, butter, honey, dates, 1 cantaro = 89.723 kg; for flax, 1 cantaro = 108.100 kg. Candia: 1 cantaro = 44 oka = 100 rottoli = 17,600 drams = 56.161 kg. Constantinople: 1 cantaro = 44 oka = 176 checky = 56.161 kg, also 1 cantaro = 450 oka = 57.438 kg. Genoa: 5 cantaro = 1 peso; 1 cantaro = 6 rubbi = 100 rottoli = 150 libre = 1,800 oncie = 52.325 kg. Mallorca (Palma): 1 cantar = 4 arroba = 104 rottoli = 41.603 kg; also used: cantaro barbaresco = 42.030 kg; 3 cantaros = 1 cargo. Minorca (Mahon): cantaro majorina = 104 rottoli = 32.1 great pounds or 104 small pounds = 41.930 kg; cantaro barbaresco = 100 rottoli = 42.030 kg. Sicily: cantaro grosso = 87.351 kg; cantaro sottile = 79.413 kg. Rome: cantaro grosso = 10 cantari sottili = 100 decini = 1,000 lira; each lira = 12 unces; 1 cantaro grosso = 339.295 kg. Rarely used was a cantaro of 150 lira and one of 250 lira. Smyrna: 1 cantaro = 7.5 batman = 45 okas = 57.818 kg. Romania (Bucharest, the Walachia): 1 cantaro = 56.726 kg.

Caractero   Pharmaceutical weight in Spain; 3 caractero = 1 obolo; 6 caractero = 1 scrupulo; 18 caractero = 1 drachma; 144 caractero = 1 oncia; 1 caractero = 0.2 gm.

Carak	Persian = 750 gm.
Carat	Gold, silver and jewel weight. Bologna: 1 carat = 4 gran = 0.108 gm; 10 carats = 1 ferlino; 160 carats = 1 oncia. Florence: 1 carat = 8 ottave = 14.15 gm. Genoa: 1 carat = 13.22 gm. Milan: 1 carat = 9.8 gm. Turin and Piemonte: 1 carat = 10.25 gm. Venice: 1 carat = 0.2 gm.
Cargo	Spanish market weight. Alicante: 1 cargo = 2.5 quintal = 10 arrobas = 240 libras majores = 360 libras menores = 124.52 kg. Mallorca and Minorca: 1 cargo = 3 quintals = 312 rottoli = 131.135 kg.
Cash	See Canton: = $\frac{1}{10}$ candarine or fen = 1/100 mace; 1 mace = 3.16 gm.
Cassico	Oil weight in Sicily = 10.196 kg.
Catti, catty	Market weight in East India, Indochina, China, Japan, and Philippines. Also called Gin. Acheen (Sumatra): 1 catti = 20 buncal = 100 tals = 200 pagodas = 1,600 mas = 6,400 capongs = 960.25 gm. Amboina: 100 catti = 1 pecul; 2,500 catti = 1 coyang rice. As gold and silver weight: 1 catti = 20 tael = 320 mas = 1,280 copangs; this catti = 590.5 gm. Batavia: 1 catti = 16 tail; 100 catti = 1 small bahar: 450 catti = 1 heavy bahar; 1 catti = 617.5 gm. In other places on Java and in Malacca 1 catti = 615 gm. Canton (China): 100 catti = 1 pecul or pic; 1 catti = 16 lyangs = 160 tschen = 1,600 swin = 16,000 lis = 604.875 gm. Japan: 1 catti = 594 gm. Manila: 22 ounces or piasters are called a catti = 594.375 gm. Singapore and Prince of Wales Island: 1 catti = 604.875 gm = 16 tails or tales; 100 catti = 1 small pecul; 300 catti = 1 basar; 400 catti = 1 coyan. Queda (Indochina): 1 catti = 735.5 gm. Siam: 1 catti = 613.6 gm.
Centigram	Metric = 1/100 gm.
Centinajo, Centi-pondium (in Milan)	In Milan since 1803 the libbra (pound) italiana or libbra metrica = 1,000 gm. Quintal = 10 rubbi = 100 libbre = 1,000 ounces = 10,000 grossi = 100,000 denari = 1,000,000 grani = 100 kg.
Centner, cwt., Centinajo, Can-tarelli, Cantaro, Quintal	Commercial weight of 100 (up to 120 pounds) used in most European countries. The weight of a pound was different in different countries. See Pound.

Chaqui, Cheky	Gold and silver weight in Basra (Asiatic Turkey) = 100 miscals or drams = 1,600 kara = 6,400 gran = 466 gm.
Charge	Oil weight in Geneva = 126.696 kg. Market weight in France = 3 quintaux = 300 livres = 146.852 kg. Holland, Brussels, Antwerp: 1 charge = 2 ballen = 4 centner = 400 pounds = 187.464 kg.
Chariot, Wag, Wog	Dutch wool weight: 2 chariots = 1 sack; 6 chariots = 1 seltier; 1 chariot = 77.353 kg.
Chattak, Chittak	A unit to weigh liquids, also grain, in Calcutta and East India. For grain 1 chattak = 42.4 gm; 5 chattacks = 1 kunke; 20 chattacks = 1 roik; 80 chattaks = 1 pallie; 1,600 chattaks = 1 soalli; 15,600 chattaks = 1 kahun. To weigh liquids: 1 chattak = 43.6 gm; 4 chattaks = 1 puah; 16 chattaks = 1 seer; 640 chattaks = 1 maon.
Chaval	India, 15 mg = $\frac{1}{8}$ rati = $\frac{1}{64}$ masha = 1/798 tola; 1 tola = 11.7 gm.
Cheh	See Canton, gold and silver weight.
Cheky, Chequi, Tscheki	Turkey = 318.667 gm; 2 chequi = 1 rottolo; 4 chequi = 1 oka = $\frac{1}{2}$ small batman = $\frac{1}{4}$ big batman; 234 chequi = 1 cantaar. The same chequi as gold and silver weight = 100 dram = 1,600 kara = 6,400 gran. Smyrna: 1 cheque = 320.875 gm.
Chouw	Surat (East India), pearl weight = 0.012 gm; 13.75 chouw = 1 retty; 330 chouw = 1 tang.
Clove, Nail, Nagel	Wool weight in England = 3.127 kg; 2 cloves = 1 stone; 4 cloves = 1 tod; 26 cloves = 1 wey; 52 cloves = 1 sack; 624 cloves = 1 last wool.
Coccio	Gold and silver weight in Sicily = 0.55 gm; 485 coccios = 1 oncia; 5,820 coccios = 1 libbra.
Cola	Aleppo (Syria) = 7 vesnos = 35 rottoli = 66.496 kg.
Commerzlast	Hamburg = $1\frac{1}{4}$ Schifflast = $2\frac{1}{2}$ tons = 5,000 Hamburg pounds = 2,422.296 kg.
Copang	Gold and silver weight in Sumatra = 0.375 gm; as commercial weight = 0.15 gm.
Coyan	See Prince of Wales Island.
Crat	Gold and silver weight in Mocha (Yemen): 16 crat = 1 cafla; 160 crat = 1 wackega; 1 crat = 0.2 gm.
Dam	East Indian = 20.96 gm.
Danich, Darchini	Arabian = 0.4 gm = $\frac{1}{10}$ darchini.
Darachnu	Islamic medical weight (drachm) = 4.25 gm.

Decagram, Gros	Metric = 10 gm.
Decigram	Metric = $\frac{1}{10}$ gm.
Decina	Commercial weight in Rome = $\frac{1}{10}$ light cantaro = 1/100 heavy cantaro = 3.393 kg.
Denaro, Denier	In Italy a coin, also a commercial weight = $\frac{1}{24}$ ounce = 24 grani. Sardinia: 1.28 gm; Florence, Leghorn, Pisa: = 1.75 gm; Lucca 1.3 gm; Parma 1.14 gm; Piacenza 1.12 gm; Rome 1.17 gm; Venice and Milan 1 gm. As gold and silver weight: 1 denaro = 24 grani = 576 granottini. Portuguese equivalent is the denheiro = 19.13 gm = $\frac{1}{12}$ marco.
Denier, Denaro	Coin and weight in Switzerland and France. France: 1 denier = $\frac{1}{3}$ gross = $\frac{1}{24}$ ounce = 1/192 troy marc; 1 denier = 2 felins = 24 grains = 1.27 gm. Lausanne: 1 denier = $\frac{1}{4}$ quart or gros = $\frac{1}{16}$ lot = $\frac{1}{32}$ ounce = 1/512 pound; 1 denier = 18 grains = 1 gm.
Derhem, Dirhem, Derham	Gold and silver weight in Persia = 2 miscal = 12 dungs = 9.36 gm. The derhem, derived from Greek drachm, is basis of Islamic weight standards; its value differed in different Islamic countries. For silver it was 2.97 gm or 2.82 gm. Standard weight of dirham in selling goods (Hinz, 1955) = 3.125 gm with deviation, e.g. in Persia = 3.2 gm.
Deusquin	Dutch = $\frac{1}{2}$ troisquin = $\frac{1}{16}$ angel = 1/320 ounce = 1/2,560 troy marc = 0.097 gm.
Dinar	Islamic = 1 mitgal = 4.23 gm.
Drachm, Drachme	Polish = $\frac{1}{4}$ lutow (lot) = $\frac{1}{8}$ ounce = 1/128 funt (pound) = 3 skrupulow = 24 granow = 132 granikow = 3.15 gm.
Drachma	Spanish pharmaceutical weight, $\frac{1}{8}$ onca = 3 scruples = 6 oboles = 12 gran. Elsewhere = 60 gran; Bern, 3.1 gm; Germany (Nuremberg drachme) 4 gm; Florence 3.57 gm; Hamburg 3.78 gm; England 3.93 gm; Holland 1 gm; Prussia 3.76 gm; Rome 3.63 gm; Venice 3.16 gm; Vienna 4.68 gm.
Dram, Drame, Dramme	Commercial weight in Turkey: Baghdad (Basra, Basora) 3.25 gm; Constantinople (dramme, drachme, dirhem) = 1/100 cheky = 1/176 rottel = 1/400 oka = 1/800 small batman = 1/3,200 big batman and 1/17,600 cantaro; 1 dirhem = 3.2 gm. As gold and silver

243

	weight 1 cheky = 100 dramme; 1 dramme = 3.1 gm = 16 karat = 64 grains. Smyrna: 180 drams = 1 rottol; 810 drams = 1 oka; 18,000 drams = 1 cantaro; 1 dram = 3.23 gm. As gold and silver weight 1 dram = 3.23 gm = 1/400 oka. Walachia: 1 dram = 3.22 gm = 1/400 occa = 1/17,600 kantar.
Droit	See Moneyers' weights.
Ducat	Austria: 1 ducat = 60 gran = 3.52 gm; 1 gran = 0.0587 gm.
Duella, Duelle	Old French apothecaries weight = $\frac{1}{3}$ ounce = 8 scruples = 192 grains = 10.2 gm.
Dung	Gold and silver weight in Persia = 0.778 gm = $\frac{1}{6}$ miscal = $\frac{1}{12}$ derhem.
Engel, Engelot, Engelsen	See Angel. Troy weight in Holland: 1 troy pound = 320 engels; 1 engel = 1.56 gm = 8 troiquins = 16 deusquins = 32 as; marc of Cologne = 152 engels.
Escrupulo	See Rio de Janeiro = 1.2 gm.
Estelin	Old French gold and silver weight = 2 gm = $\frac{1}{20}$ ounce = 1/160 troy mark = 2 mailles = 4 felins = 28.8 grains.
Fanoe, Fanon	Gold and silver weight in East India = 0.378 gm; $11\frac{1}{2}$ fanoes = 1 miscal in Calcutta.
Farasila, Frassola	Islamic = $\frac{1}{20}$ bahar.
Fatil	Islamic = 1/432 gon = 0.045 gm.
Felin	Old French troy weight = 0.38 gm = $\frac{1}{2}$ maille; $3\frac{1}{3}$ felin = 1 denier; 4 felin = 1 estelin; 10 felin = 1 gros; 80 felin = 1 once; 640 felin = 1 troy marc.
Ferlino	Italian: Bologna 1.6 gm; Ferrara 1.73 gm; Modena 1.78 gm; 1 ferlino = 10 karat = 40 grains.
Fetr	Medieval Persian = $\frac{1}{10}$ harwar = 8.33 kg.
Föring	Commercial weight in Danish islands, Iceland = 10 pounds = 4994 gm.
Fouang, Foang	Gold and silver weight in Siam = $\frac{1}{8}$ tical = $\frac{1}{32}$ tael = 1.875 gm.
Fuen, Fen, Fou, Swin, Twen	Chinese gold and silver weight. As gold weight 1 fu = $\frac{1}{10}$ see = 1/100 hoa = 1/1,00 li = 1/10,000 fuen = 10 tschin = 100 jai = 1,000 miao = 10,000 mo = 100,000 tsiun = 1,000,000 sun. As silver weight 1 fen or fuen or swin = 10 li = $\frac{1}{10}$ tsien or tschen = 1/100 leang or lyang = 1/1,600 catti = 1/160,000 pic or pecul; 1 fuen = 0.378 gm.

Fun	See Japan.
Funda, Funta	Russian pound as gold and silver weight. = 96 solotniks = 409.16 gm.
Funt, Pfund	In Polish Kingdom = 16 uncyi = 32 lutow = 128 drachms = 384 scruplow = 9,216 granow = 404.6 gm = $\frac{1}{25}$ kamienec (stone) = 1/100 centner.
Gändom	Persian grain of wheat = 0.048 gm.
Gantang	See Sumatra.
Gantas	Commercial weight in Queda (Malacca) = $\frac{1}{16}$ hali = $1\frac{15}{16}$ pound.
Garme, Kermet	Arabic = 3 ounces.
Garse, Garsa	East India, for grain or rice = 24 French charges or 72 quintaux = 3,522.54 kg = 68.5 Prussian centners.
Gaunting	Java, for rice = 6,916 gm.
Gauza	Islamic: common gauza = 29.75 gm, the royal gauza = 25.5 gm.
Gedang	Weight used on the Spice Islands for pepper = 1,976 gm.
Gin	See Catty.
Glied, Stein	Wool weight in Fulda (Germany) = 21 Fulda-pounds = $\frac{1}{5}$ wollcentner = 10.719 kg.
Gon	Persian grain = 0.045 gm; later, 0.048 gm.
Grän	As gold and silver weight: Cologne as the gold standard mark = 24 karat; 1 karat = 12 grän. Silver standard mark = 16 loth; 1 loth = 18 grän. In both cases 1 Cologne mark = 288 grän; 1 grän = 0.82 gm. Alexandria: 1 grän = $\frac{1}{4}$ quirat = $\frac{1}{64}$ drachme = 0.05 gm. Antwerp: gold standard karat = 12 grän; silver standard = 24 grän = 1 pfennig; 288 grän = 1 mark; 1 grän = 0.86 gm. Augsburg: 1 grän = 0.82 gm. Berlin, same as Cologne. Bern: gold standard karat = 32 grän; silver standard: 18 grän = 1 loth; 24 grän = 1 denier; 288 grän = 1 mark; 1 gold grän = 0.32 gm; 1 silver grän = 0.852 gm. Denmark: 1 grän = 0.825 gm. Nuremberg 0.9 gm; Riga 0.72 gm; Sweden 0.76 gm. Vienna: 1 grän (Mändel Gewichtsgrän) = 0.59 gm; 60 grän = 1 ducat; 301.5 grän = 1 loth and 4,824 = 1 mark.
Grain, Gran	In England = $\frac{1}{24}$ pennyweight = 1/480 ounce = 1/5,700 troy pound = 0.065 gm. In avoirdupois standard 437.5 grains = 1 ounce; 7,000 grains = 1 avdp.

	pound. Old Scotch troy pound = 7,620 grains. In France troy mark = 4,608 grains; in grains once = 587, gros = 72, denier = 24, maille = 14.4, felin = 7.2; grain = 0.53 gm.
Gram	The unit of the metric-decimal system.
Gran, Grän, Grano	Pharmaceutical weight units used in medical prescriptions; 20 gran = 1 skrupel, 60 = 1 drachme, 480 = 1 ounce, 5,760 = 1 medicinal pound. In most parts of Germany the gran weight of Nuremberg was generally used. The weight of 1 medicinal gran in Holland = 0.067 gm; Berlin 0.05 gm; Bern 0.05 gm; Denmark 0.067 gm; Munich 0.062 gm; Nuremberg 0.067 gm; Sweden 0.063 gm; Vienna 0.0746 gm.
Granikow	Poland, 5.5 = 1 granow = 0.043 gm. See Funt.
Granottino	Smallest gold and silver weight in Turin (Piemonte) = $\frac{1}{24}$ grano = 1/576 denar = 0.002 gm.
Granow	Polish (see Funt, and Granikow). Russian, for hay = 20 Russian pounds = 8,183 gm.
Gros	French = 3 deniers = 72 grains = 3.9 gm = $\frac{1}{8}$ once. Lausanne: 8 gros = 1 ounce; 128 gros = 1 livre; 1 gros = 4 deniers = 72 grains = 3.9 gm.
Gunja, Krishnala	Basic weight in Madras; a seed weighing 65–130 mg.
Gyrath	See Carat, also Algiers.
Habba	Arabian gran weight = 0.05 gm; in Egypt (19th century) 0.065 gm.
Hali, Nali	Commercial weight in Queda (Malacca) = 16 gantas.
Hardal	Islamic (a mustard grain) = ca. 0.00071 gm.
Harruba	Islamic = 1 girat = $\frac{1}{24}$ mitgal = 0.195 gm (grain of St. John's bread).
Harsela	Commercial weight in Egypt = 400 drams = 1,276.4 gm.
Harwar	Islamic, load of an ass. In Persia 288, later, 300 kg.
Hectogram	Metric = 100 gram.
Heller	Gold and silver weight = 0.45 gm = $\frac{1}{2}$ pfennig = $\frac{1}{8}$ quent = $\frac{1}{32}$ loth = $\frac{1}{64}$ ounce = 1/512 mark.
Himl	Islamic; a camel load = 300 männ = 600 ratl = ca. 250 kg.
Hali	See Gantas.
Harwar	See Fetr.
Hoa	Chinese, see Fuen.
Hoon	See Canton.

246

Hoot	See Canton.
Hüttencentner	A hundredweight used in mines in Germany, 10–20 pounds heavier than regular centner.
Hundert, Centner, Zentner	Equals $\frac{1}{20}$ of an English ton = 4 quarters or 112 pounds avoirdupois = 50.797 kg.
Istar	Islamic = 20 gm (the Greek stater).
Jai	Chinese gold weight = 10 miao = 100 mo = 1,000 tsiun = 10,000 sun = $\frac{1}{10}$ tschin = 1/100 fou = 1/1,000 see = 1/10,000 hoa = 1/100,000 li; 1 li = 0.034 gm.
Kahun	Weight for grain in East India = 1,318.17 kg = 2 candils = 40 maons = 16 soalli = 320 pallies = 128 roiks = 5,120 kinkes = 25,600 chattaks. The gram weight of the Kahun is different in different provinces.
Kalvar	Persian = 100 batman. Kalvar of Miranda = 594.2 kg. Big kalvar of Tauris = 464.2 kg. Small kalvar of Tauris = 436.3 kg.
Kantar	See Cantar.
Kara, Karat, Taim	Gold and silver weight in Turkey, 0.2 gm = $\frac{1}{16}$ dram = 1/1,600 cheky (see carate) = 4 grains.
Karat, Carate, Quilat, Kara	Gold and silver weight: 1 mark gold = 24 karat; 1 karat = 12 grän. According to the mark weights the karat weight changes in different countries: in Berlin 10.62 gm; Bern 10.2 gm; Cologne 10.62 gm; Leipzig 10.62 gm. As jewel and pearl weight: 1 karat = 4 grän: Amsterdam 0.207 gm; Batavia and East India 0.198 gm; Berlin 0.206 gm; Frankfurt a/M 0.208 gm; Hamburg 0.206 gm; London 0.209 gm; Leghorn 0.198 gm; Vienna and Austria 0.208 gm.
Kargo, Chargo, Kariko	See Cargo.
Karsch	See Vienna.
Katje	See Sumatra.
Katti, Kätti	See Catti. Chinese = 604.79 gm; also 601.28 gm.
Ketti	See Catti.
Khahoon	See Calcutta.
Kilogram	Metric = 1,000 gm.
Kirath	Arabic = 2 kestufs = 4 gran = 8 kepath.
Kleud, Klender, Stein	Wool weight in Haunau (Germany) = 18 woolpounds = 10.516 kg; 5 kleud = 1 centner of wool.
Koonke	See Calcutta.

Korrel	Smallest Dutch weight = $\frac{1}{10}$ wigtje = 1/100 load; 1/1,000 once = 0.1 gm.
Koyang	Batavia = 27 pecul = 9 small or 6 big bahars = 1,667 kg.
Krinne	Bünden (Switzerland); big krinne = 693.8 gm = $\frac{1}{75}$ centner; 12.5 krinne = 1 rupp; small krinne = 520.38 gm.
Krishnala	See Gunja.
Kulack	Weight for grain and rice in Batavia = 4.304 kg.
Kunke	Weight for grain in Calcutta = 5 chattak = $\frac{1}{4}$ roik = $\frac{1}{16}$ pallie = 1/320 soalli = 1/5,120 kahun = 281.92 gm.
Lägel	Weight for steel in Austria = $\frac{1}{2}$ saum = 70.015 kg.
Lana	Russian, = $\frac{1}{2}$ pound = 8 solotnik = 34 gm.
Last	Commercial weight in shipping, very different in different countries and in same country for different goods.
Liang, Tehl	Chinese, between 32 and 39 gm.
Libbra, Lira	Italian pound of different weights in different cities: Bologna = $\frac{1}{25}$ peso = 12 oncie = 192 ferlini = 1,920 carati = 7,680 grains = 363.13 gm. Naples = 321 gm; Rome = 339.3 gm; etc. Libbra in Bologna = 342.7 gm. In Brazil 458.8 gm; Portugal 459 gm.
Liespound, Liespfund, Lispound	Commercial weight in North Germany, = 14 pounds in Bremen, Hamburg, Lübeck, Rostock; and other towns. Denmark 16 pounds; Sweden, Riga, and Reval 20 pounds. According to area, between 4.5 and nearly 8 kg.
Lira	See Libbra.
Littre, Litre	Commercial weight in the Walachai = 100 drams = $\frac{1}{4}$ oka = 1/176 kantar = 322.3 gm.
Livre (poid de marc)	France, pre-metric pound = 16 ounces = 128 gros = 384 deniers = 9,216 grains = 489.5 gm. For silk only 15 ounces = 459 gm. In the early 19th century livre was also used for $\frac{1}{2}$ kg = 500 gm.
Load	English; 1 load hay = 2,160 pound = 979.473 kg; 1 load wool = 12 sacks = 24 weys = 156 tods = 312 stone = 624 cloves or nails = 1,980.713 kg.
Lodra	Islamic = 564.4 gm.
Lof	Riga = 41.814 kg.
Lood, Loode, Loth	Dutch commercial weight = $\frac{1}{10}$ once = 1/100 pond = 10 wigtje = 100 korrel = 10 gm. Old Dutch troy weight: 2 loods = 1 once, 1 lood = $\frac{1}{16}$ mark = $\frac{1}{32}$ pound = 16 engel = 320 *as* = 15.4 gm.

Lot, Loth	Commercial weight, Germany, Austria, Switzerland, Denmark, Sweden, etc., = $\frac{1}{32}$ pound. Zürich, e.g., had also a heavy pound of 36 loth; 1 loth = $\frac{1}{2}$ ounce = $\frac{1}{16}$ marc = 4 quentchen = 16 pfennig = 32 heller = 4,096 richtpfennig. Its weight in grams depends on the weight of the pound in different countries.
Lutow, Loth	Polish, = $\frac{1}{2}$ ounce = 12.67 gm.
Lyang	See Canton; 37.78 gm.
Maille	French gold and silver weight = 2 felins = 14.4 grains = 0.78 gm = $\frac{1}{5}$ gross = $\frac{1}{40}$ ounce = 1/320 troy marc.
Maj	Apothecaries' weight. See Japan.
Mangal, Mangalis, Mangelin	Weight of pearls and jewels at Coromandel = 0.25 gm.
Männ	The ancient mina, = 2 ratl = 130 dirham. In Egypt 819 gm; about its local and historical changes see Hinz (pp. 17 ff.); it went up to about 6 kg.
Maon	Commercial weight in East India. In Calcutta, 1 maon (pucca) = 40 seers = 100 paloins = $\frac{1}{20}$ candil = 10.923 kg. For grain 1 maon = $\frac{1}{40}$ kahun. Maon in Bombay = 12.70 kg = $\frac{1}{20}$ candy. Maon in Madras = 11.340 kg. Maon in Mocha = 13.237 kg.
Marca	In Spain the Aragon marca = $\frac{1}{2}$ pound = 8 oncas = 32 quartos = 128 arienzes (adarmes) = 4,096 granos = 230 gm. The Castilian marca (Madrid) = 8 onças = 64 ochavos = 128 adarmes = 384 tomines; or 1 marca = 24 karat = 96 gran gold or 12 dineros or 288 gran silver; or as pharmaceutical weight = 8 onças = 64 drachmos or escrupelos or 364 oboles or 1,152 caracteres = 4,608 granos = 230.45 gm. In Barcelona the Catalanian marca = 272.67 gm. The marca of Valencia = 237.9 gm. In Italy: Milano, Ferrara, Bergamo, 1 marca = 8 oncie = 192 denari = 4,608 grani = 235 gm. The marca of Piemonte = 246 gm. The marca of Venice = 8 oncie = 32 quarti = 192 denari = 1,152 carati = 4,608 grani = 238.5 gm.
Marco	Brazil and Portugal = 8 onças = 64 octavas = 192 escrupelos = 4,608 granos = 299.5 gm = $\frac{1}{2}$ aratel (pound).
Mark, Marc, Marck	Old German weight = $\frac{1}{2}$ pound. In Denmark = 8 unzen = 16 loth = 64 quent = 256 ort = 249.67 gm.

249

In Germany, the Cologne mark was most generally accepted = 233.75 gm. Weight in other countries slightly different (see Troy mark).

Masha, Massa, Mas, Mahs, Mace   Gold and silver weight in East India = $\frac{1}{12}$ tola in Surat but $\frac{1}{15}$ tola in Bombay = 1/120 seer = 1.15 gm. For jewels or pearls, 1 massa = $\frac{1}{20}$ ruttee = 1/480 tang = 0.045 gm. As commercial weight in Sumatra 1 massa = 4 copang = $\frac{1}{8}$ pagodas = $\frac{1}{16}$ tael = $\frac{1}{80}$ bunkal = 1/1,600 catties = 10.6 gm; 320,000 mas = 1 bahar. In the sixteenth century (Hinz, p. 23) = 1.0042 gm. According to Luther (1961) 1 masha = 972 mg.

Maund   Prince of Wales Island and Singapore, weight for rice = 82 English pounds = 37.2 kg = $\frac{1}{2}$ sack. In Sumatra 1 maund of rice = 20 bamboos = 75 English pounds = 34.01 kg. India: 1 maund = 40 seer = 37.2 kg; see also Bengal.

Me, Monme   Japan, 1 me = 1/1,000 kwan = 3.75 gm. Sumatra, 1 me = $\frac{1}{8}$ pagoda = 0.6 gm.

Metska, Meteka, Mitiga, Methkal   Gold and silver weight in Turkey and North Africa. In Algiers 1 metekal = 4.67 gm; in Egypt 4.68 gm; in Syria and Tripoli 4.8 gm.

Mian   Gold and silver weight in Malacca, Singapore = 2.67 gm. See also Prince of Wales Island.

Miao   Gold weight in China = 10 mo = 100 tsiun = 1,000 sun = $\frac{1}{10}$ jai = 100 tschin = 1,000 fu = 10,000 se = 100,000 hao = 1,000,000 li; 1 li = 0.05 gm.

Migliago   See Rome.

Migr   Egypt = 18 qirat = 3.51 gm.

Migräb   Anatolia = 750 gm.

Millier   Metric = 1,000 kg. See Paris.

Milligram   Metric = 1/1,000 gm.

Mina, Mna   Commercial weight in Alexandria = 755.67 gm; in Cairo = 596.2 gm; in Syria = 589.33 gm.

Mite, Heller   English medical and moneyers' weight (see there) = $\frac{1}{20}$ grain = 1/480 pennyweight = 1/9,600 ounce = 0.015 gm.

Mitqal   Islamic, derived from the Roman solidus = $1\frac{1}{2}$ (exactly 10/7 dirham) = 4.464 gm.

Moneyers' weight   England, 1 grain = 20 mites; 1 mite = 24 droits = 480 periots; 1 periot = 24 blancs; 1 grain = 0.065 gm.

Monme   See Me, Japan.

Mozetta	Salt weight in Corfu and Paro (Ionian Isles) = $\frac{1}{2}$ sacco = $\frac{1}{60}$ centinajo = 995.1 gm.
Myriagram	Metric = 10 kg.
Nail, Nagel, Näl	Wool weight in England. See Clove. In Brussels, Antwerp, and elsewhere in Holland, 27.5 näl = 1 chariot; 1 näl = $\frac{1}{55}$ sack = 1/330 seltier = 6 pounds = 2.812 kg.
Naqir	Islamic = 0.045 gm.
Nass	Arabic (Mecca) = $\frac{1}{2}$ ugiya = 62.5 gm.
Nawa	Arabic = 15.6 gm.
Nelli	Rice weight in Sumatra = 14.595 kg.
Nohod	Persian (pea) = $\frac{1}{24}$ mitgal = 0.18 gm; after the fifteenth century, 0.195 gm.
Nüge	In medieval Anatolia two weight standards under this name: one = 641.4 gm, the other = 250.1 gm.
Occa	Weight in the Walachia = $\frac{1}{44}$ kantar = 1.289 kg. See Oka.
Ochava	See Madrid.
Oertchen	See Ort.
Ogga	Osmanic = 400 dirham = 1.283 kg.
Oitavo	See Lisbon.
Oka	Turkey, 44 big oka = $58\frac{1}{3}$ small oka = 1 cantaro (see Cantaro). Small oka = 2 rottel = 4 checky = 400 dirhem = 1.249 kg.
Ounce, Oncia, Onca, Once, Onça, Unze	See Pound and Troy pound. One pound in Roman times and the later medicinal pound = 12 ounces. The commercial modern pound usually = 16 ounces. Weight in grams depends on local weight of pound. In France 1 ounce = 8 gros = 20 estelins = 24 deniers = 40 mailles = 80 felins = 576 grains; in medical weight 1 ounce = 8 drams = 24 scruples = 576 grains = 30.6 gm. French law of 1800 set the ounce at 100 gm (1 hectogram) and the gros at 10 gm (1 decagram); the denier = 1 gm and the grain = 0.1 gm (1 decigram). In 1812 weight of 1 pound was set at 500 gm or $\frac{1}{2}$ kilogram; the mark at 250 gm; ounce at $\frac{1}{16}$ pound or $\frac{1}{8}$ mark = 31.25 gm. This became definite law in 1825. In Holland 1 ounce of old troy weight = 20 engels = 160 troiquins = 320 deusquins = 640 as = 30.75 gm. The once = $\frac{1}{10}$ pond (kilogram) = 10 looden = 100 wigtjes (grams) = 1,000 korrels. In medical weights 1

251

once = $\frac{1}{12}$ pond = 375 wigtjes = 8 drachms = 24 scruples = 480 gran = 31.25 gm. Switzerland: in Geneva 1 ounce = $\frac{1}{15}$ light pound = $\frac{1}{16}$ of a medium pound and $\frac{1}{18}$ of a heavy pound = 24 deniers = 576 grains. In Lausanne since 1822 the French ounce (see above) accepted by law = 31.25 gm. In Italy the oncia has very different weights; in Venice and Verona the light oncia = 25.2 gm; in Padua the heavy oncia = 36 gm. In Milan since 1803 the oncia = 100 gm. In England, by the commercial avoirdupois standard, 16 ounces = 1 pound, 448 ounces = 1 quarter, and 1,792 ounces = 1 hundred; 1 ounce = 16 drams = 27.2 gm. By the imperial troy or coinage standard 1 ounce = $\frac{1}{12}$ pound = 20 pennyweight = 480 gran = 9,600 mites = 31.08 gm.

Øre
Norway, Sweden, Iceland. Equivalent to the old Roman ounce; 1 øre = 7 drachmer = 28 scripula = 26.805 gm.

Ort
Denmark = $\frac{1}{4}$ quent = $\frac{1}{16}$ loth = $\frac{1}{32}$ ounce = 1/256 mark = 1/512 pound = 0.975 gm.

Ottava, Octava, Outava
Portugal and Brazil, 1 ottava as gold and silver weight = $\frac{1}{8}$ ounce = $\frac{1}{64}$ marco = 1/128 aratel (pound) = 3 escrupulos = 72 granos = 3.6 gm. Italy: in Genoa 1 ottava = $\frac{1}{8}$ carato = 1/192 libra scarsa = 1.7 gm. In Florence 1 ottava = 1.78 gm. As weight for jewels in Portugal and Brazil 1 ottava = $\frac{1}{8}$ onça = 3 escrupulos = 9 quilates = 36 granos = 3.6 gm.

Pack, Bale
English commercial weight = 240 pounds wool = 108.872 kg; 1 pack of yarn = 1.814 kg = 4 English pounds.

Packen
Russian = 3 berkowitz = 30 pud = 1,200 funt (pounds) = 490.790 kg.

Pagoda
India, Sumatra, etc., 200 pagodas = 1 catti = 960.2 gm. In Madras since 1950 = 54.79 grains = 3.55 gm.

Pala
Madras = 320 krishnala.

Palie, Pallie, Paloin
East India: in Calcutta = $\frac{1}{8}$ mahon = $\frac{1}{20}$ soalli = 1/160 candil = 1/320 kahun = 4 roiks = 16 kunkes = 80 chataks. In Bombay 1 palie = $\frac{1}{20}$ pherra = 1.701 kg; see also Maon.

Parah
Weight for grain and rice in East India. One type = $\frac{1}{8}$ candy = 16 adowlies = 64 seers = 128 tiprees =

20.322 kg. The other type $= \frac{1}{25}$ candy $= 1/100$ morah $= 20$ adowlies $= 150$ seers $= 300$ tiprees $= 47.621$ kg. In Sumatra as weight for salt 1 parah $= 26$ punihs $= 47.435$ kg.

Parto   Naples, gold weight $= 1/100$ denaro $= 0.011$ gm.

Passirgewicht   The weight of a gold coin still acceptable, though it is not exactly the official standard weight of this coin.

Pecul, Pekul, Pic, Picol, Picul   East Indian, $= 100$ catties $= 1,600$ täls (leangs, tails) $= 59.052$ kg. See Bahar, Catti.

Pennyweight   See Ounce: England.

Percoiwitz, Perkowitz   See Berkowitz; Packen.

Periot   See Moneyers' weights.

Perma   Russian weight for hay, $= 240$ pud $= 900$ Russian pounds $= 3,926.4$ kg.

Pesau   Old French weight for chestnuts, $= 125$ to 130 pounds

Peso   Italian: one must usually distinguish between gran peso or peso grosso, the heavyweight, and peso scarso or peso sottile, the lightweight; in Bologna, 1 peso $= 9.053$ kg; Genoa, 1 peso grosso $= 26.162$ kg; Venice, 1 peso grosso $= 1\frac{7}{12}$ peso sottile $= 477$ gm; 1 peso sottile $= 301.25$ gm.

Pfennig   As commercial weight $= \frac{1}{4}$ quent $= \frac{1}{16}$ loth $= 1/512$ pound. In Bern 1 pfennig $= 1$ gm; Oldenburg 0.95 gm; Hanover (Osnabrück) 0.48 gm. As gold and silver weight 1 pfennig or richtpfennig (denarius, directorius) $= \frac{1}{4}$ quent $= \frac{1}{18}$ loth $= \frac{1}{32}$ ounce $= 1/256$ marc $= 2$ heller $= 256$ richtpfennigteile. In Berlin $= 0.9$ gm; Bern 0.95 gm; Mainz 0.96 gm; Nuremberg 0.92 gm; Zürich 0.8 gm; Vienna 1.1 gm.

Pfund (pound)   See Aratel, Cargo, Funt, Libbra, Lira, Livre, Pond, Pud, Rotolo. In Roman times the old Oscan pound $= 273$ gm; in Imperial period $= 327.45$ gm. The uncia (ounce) $= \frac{1}{12}$ pound $= 27.29$ gm $= 4$ sicilici each to 6.82 gm; 1 sicilicus $= 6$ scripula, each $= 1.137$ gm. In modern times as commercial weight 1 pound $= 16$ ounces $= 32$ loth $= 128$ quent. There were often a light and a heavy pound. The medical pound $= 12$ ounces; in some places in Germany and Switzerland pounds went up to 36 and 40 loth; 20 to 25 pounds $= 1$

	stein (stone); 100 to 120 pounds = 1 centner. See also Liespound and Ship's pound. The actual weight of 1 pound in grams was different in most places (see Niemann or Nelkenbrecher and many others from the first half of the nineteenth century). Frachtpfund in Germany = 300 pounds.
Pherra	East Indian weight for rice = 20 palies = 34.015 kg.
Piastre	Coin, also a weight used in Philippines, = 1 ounce = $\frac{1}{10}$ tole gold = $\frac{1}{11}$ tole silk = 27 gm. See also Morocco.
Pice, Puah	India, for liquids = $\frac{1}{4}$ seer = 1/160 maon = 4 chattak = 68.3 gm.
Picol, Tan	Chinese = 60.128 kg; also 45.36 kg.
Pisc	Central Africa = 0.67 seron = 2 aquiraques = 4 mediatablas = 8 gm.
Pond	Dutch = 10 onces = 100 looden = 1,000 sigtjes = 10,000 karrels = 1,000 gm.
Pound	English; see London.
Puah	See Pice.
Pucca	See Maon.
Pud	Russian commercial weight = 40 Russian pounds = $\frac{1}{10}$ berkowitz = 16.360 kg.
Pugillus	In pharmaceutical prescriptions a pinch or as much as can be taken with three fingers.
Punkho	See Calcutta.
Qamha	Islamic grain of wheat = 0.0488 gm.
Qitmir	Islamic = 0.045 gm.
Quadrans	In medicine, 3 ounces or $\frac{1}{4}$ medical pound.
Quardeel, Quarteel	Thrangewicht (blubber weight); in Hamburg = 4 centner = 217 kg.
Quart	A weight for grain, in Bremen = 710.5 gm.
Quartano, Corta	Oil weight in the Balearics (Mallorca, Minorca) = 9 rotoli = 3.782 kg.
Quent, Quentchen, Quentlein, Quint	Commercial weight in Germany, Austria, Switzerland, Denmark, Sweden, etc. = $\frac{1}{4}$ loth = 1/128 pound; weight in grams depends on weight of local pound. In gold and silver weight 1 quent = 4 pfennig = 8 heller. See Mark. In Cologne, Berlin, Frankfurt = 3.654 gm; Copenhagen 3.673 gm; Leipzig 3.651 gm; Nuremberg 3.730 gm; Vienna 4.870 gm. In pharmaceutical weights 1 quent = 1 dram.

Quilado, Quilat, Karat	Jewelers' weight in Portugal and Brazil: $\frac{1}{3}$ escrupulo = $\frac{1}{9}$ octava = $\frac{1}{72}$ onça = 0.2 gm. In Spain 1 quilado = 1/140 onça = 0.2 gm = 4 granos.
Quintal, Quintar	See Centner, Cantaro, Centinajo. One quintal = 4. arrobas = 128 libras. In Brazil and Portugal 4 arrobas = 128 libras = 58.745 kg. In France 1 quintal = 100 livres = $\frac{1}{3}$ charge = 48.951 kg. In the metric system in France 1 quintal = 100 kg. In Switzerland and Geneva 1 quintal of oil or brandy = 104 pounds = 77.086 kg. Quintals of different weight were used in the different provinces of Spain. In Mexico the quintal = 46.072 kg. Islamic, 1 quintar = 100 ratl, occasionally also 100 männ. As gold weight = 42.33 kg. (See Hinz re quintas used in Islamic countries; pp. 24 ff.).
Quintas	North Africa = 7.06 gm.
Quirat, Quilat, Karat	Gold and silver weight in Asia Minor, Egypt, Mecca, Syria = $\frac{1}{16}$ drachm = $\frac{1}{24}$ mitqal = $\frac{1}{43}$ mas = 1/175 pärdan = 0.2004 gm. Islamic = $\frac{1}{20}$ mitqal = 0.223 gm.
Rachim	Gold and silver weight in Sumatra = 0.063 gm.
Raik	See Calcutta.
Ratel, Rottol	Commercial weight in Persia = 382.75 gm = $\frac{1}{30}$ buttima. See Rotolo.
Rati	See Retty.
Räy	Teheran = 12 kg.
Real	Gold and silver weight in Dutch East Indies: in Java and Batavia 2.5 real = 1 tail, 9 real = 1 troy marc; 1 real = 48 stüwer = 27.33 gm.
Retty, Packaretty, Rati, Rootee	East India: in Bombay, Surat, and elsewhere 1 retty = $\frac{1}{3}$ waal = $\frac{1}{96}$ tola = 0.13 gm; in Calcutta = $\frac{1}{8}$ massa = $\frac{1}{72}$ tola gold = $\frac{1}{84}$ tola silver = 4 nely = 0.14 gm. For jewels 1 pakkaretty = 0.175 gm. One reti = 121 mg (Luther, 1961).
Richtpfennig	See Quentchen. One richtpfennig = $\frac{1}{4}$ quentlein = 0.97 gm = 256 richtpfennigteile.
Richtpfennigteil	See Pfennig = 1/256 richtpfennig = 0.0036 gm.
Rin	See Japan.
Rjo	Apothecaries' weight. See Japan.
Roba	See Arroba.
Roik	Weight in Bengal East India for grain and rice = $\frac{1}{4}$ pallie = $\frac{1}{32}$ maon = $\frac{1}{80}$ soallie = 1/1,280 kahun = 1.126 kg. See Kunke.

Rotolo, Rottolo, Rotel, Rottel — Weight in southern Italy, Turkey, North Africa, Malta, Mallorca, Minorca, etc. (For medieval value of the rotolo in Islamic countries see Hinz, pp. 28 ff.). In Turkey = $\frac{1}{2}$ small oka = $\frac{1}{4}$ small batman = $\frac{1}{16}$ big batman = 1/117 quintal or cantaro = 637.8 gm. Another rottel in use was 1/100 cantaro or 564.5 gm. In Egypt different rottoli have been used; the rotolo forforo, the most common = 423.9 gm = 1/100 cantaro. For different merchandise the number of rottoli in 1 cantaro was different: e.g. tin and mercury 102, coffee 105, ivory 110, almonds 115, gum Arabic 133. Rotolo mina = 757 gm; rotolo zaidino = 605.5 gm; rotolo zaro or zauro = 938.6 gm. In Arabia (Mecca, Medina, etc.) the rotolo = 462.8 gm; Candia and Crete 561.5 gm (the light rotoli = 341.9 gm). In Cyprus the rottolo = 12 ounces = 2.393 kg; in Rhodes 2.392 kg; in Syria 2.280 gm, but the rotolo of silk = 2.216 kg; the rottolo of Damascus = 1.900 kg = $\frac{1}{5}$ vesno = $\frac{1}{35}$ cola = 1/100 cantar. Another type of rotolo was only 1.863 kg. The rotolo in Tripoli = 1.816 gm. Africa: 1 rottlo in Abyssinia = 12 vakeas (ounces) = 144 drachms = 311 gm; Algiers = 16 ounces = 540.5 gm; Guinea 452.6 gm; Morocco 470.25 gm; Tetuan 709 gm; Tripoli 508.5 gm; Tunis 494.75 gm. In Mallorca and Minorca 1 rottolo = $\frac{1}{26}$ arroba = 1/100 cantaro barbaresco = 1/104 cantaro majorina = 420.5 gm. In Malta the light rottolo = 791.67 gm, the heavy rottolo = 870.75 gm. In the Kingdom of Naples 1 rottolo = 891 gm. In Sicily the light rottolo (rottolo sottile) = 794 gm, the heavy (rottolo grosso) = 873.33 gm.

Rubbo, Rubo, Rubbio — Italian commercial weight = 25 pounds; according to metric system = 10 libbre; by this system in Milan 1 rubbo = 10 libbre = 100 oncie = 1,000 grossi = 10,000 denari = 100,000 grani = 1,000 gm. The oil-rubbo of old standard = 25 heavy pounds (libbre peso grosso) = 1,000 quart = 700 ounces = 18.830 kg. Another oil rubbo = 25 pounds = 800 ounces = 21.284 kg. The rubbo in Parma = 8.161 kg; Turin 9,225 gm; Nice 7.750 kg; Parma 8.161 kg.

Ruttee — Gold weight in Bengal = $\frac{1}{8}$ massa = $\frac{1}{80}$ sicca = 0.187 gm. See also Surat.

Ruzma　　　　　　See Ballen. Islamic silk weight = 24.3 kg.

Sack　　　　　　　In Bombay rice is sold by sacks; 1 sack = 6 maons = 76.198 kg; in London 1 last wool = 12 sacks; 1 sack = 2 weys = 13 tods = 26 stone = 52 cloves or nails = 165.09 kg. In Holland 1 sack wool = 2 chariots = 55 nails = 154.686 kg.

Sago　　　　　　　Venetian = $\frac{1}{6}$ oncia = $\frac{1}{12}$ peso = 4.17 gm (peso sottile) or 6.63 gm (gran peso).

Sairo　　　　　　 Persian grain weight = 0.05 gm.

Satin　　　　　　　See Zethim.

Saum　　　　　　　In certain parts of Austria: Tyrol (Botzen), 1 saum = 4 centner = 200.384 kg; Vienna, 1 saum = 2.75 center = 275 pounds = 154 kg.

Saumna　　　　　 Islamic = $\frac{1}{4}$ baquila = 0.585 gm; probably also one of 1.7 gm.

Schaff　　　　　　 In Appenzell (Switzerland), 1 schaff butter = 18 pounds 25 loth Viennese weight = 10.53 kg.

Schiffslast　　　　 See Last. In Prussia = 4,000 pounds; in Hamburg 1 schiffslast = 4,000 pounds; 1 commerzlast = 5,000 pounds.

Scruple, Scrupolo　Pharmaceutical weight = 20 grains = $\frac{1}{3}$ dram = $\frac{1}{24}$ ounce = 1/288 pound. Scruple in Berlin = 1.219 gm; Nuremberg 1.331 gm; Vienna 1.458 gm; Holland 1.284 gm (since 1816 1.305 gm). In Italy the scrupolo = $\frac{1}{3}$ dramma = $\frac{1}{24}$ oncia = 1/288 libbra = 24 grani. After introduction of metric = decadic system 10 scrupoli or denari = 1 grosso; 100 scrupoli = 1 oncia; 1,000 scrupoli = 1 libbra peso medicinale; 1 scrupolo in Bologna and Florence = 1.175 gm; Milan 1.458 gm; Rome and Papal State 1.18 gm; Turin 1.069 gm; Venice 1.046 gm.

Secchia, Secchi, Secchino　Weight of salt in Corfu and Paxo; 30 secchi or sacchi = 1 centinaio; 1 secchia = 2 mozette = 1.990 kg.

See　　　　　　　 Chinese gold weight = 10 fou = 100 tschin = 1,000 jon = 10,000 miao = 100,000 mo = 1,000,000 tsiun = 10 million sun = $\frac{1}{10}$ hoa; 1 hoa = $\frac{1}{10}$ li = 0.005 gm.

Seer, Seira, Ser, Ceer, Keer, Kair　East India: 1 seer grain = $\frac{1}{40}$ maund = 2 tipres = $\frac{1}{4}$ adowlie = $\frac{1}{8}$ para = $\frac{1}{64}$ candi = 30 pices = 317.5 gm. In Persia = $\frac{1}{40}$ männ = 72.24 gm. Since 1935 in Iran = 75 gm.

Seir, Seira　　　　 See Seer. East India = 278 gm; in Persia = 303 gm.

Seltier, Serpelier	Wool weight at Antwerp, Brussels, and places in Holland; 1 seltier = 3 sacks = 6 chariots = 330 nails = 464.06 kg.
Seron	Weight in inner Africa = $1\frac{1}{2}$ piso = 2 quintas = 3 aquiraques = 4 mediatablas = 12 gm.
Set	Grain weight in Siam = 1.461 kg.
Shagvray	Persia = 10.863 kg.
Ship's pound Schiffpfund	A commercial weight in North Germany, Holland, Denmark, Russia (see Berkowitz), and Sweden (see there); its standard locally set, usually = 300 local pounds. In some places less than 300 pounds (see Berlin).
Shipping pound	See Berlin.
Shipping ton	See Berlin.
Sicca, Sicca-Rupee	Gold weight in Bengal = 10 massas = 80 ruttees = 15 gm; in Calcutta 11.6 gm.
Siqt	Silk weight equal to ruzma; Islamic = 24.3 kg.
Siranca	Islamic, ca. 166.67 kg.
Solotnik	Russian = $\frac{1}{3}$ loth = $\frac{1}{96}$ funt (pound) = 1/3840 pud = 4.25 gm. As gold and silver weight = 64 gran.
Sompi	Gold and silver weight in Madagascar = 3.6 gm.
Sorh	Indian, sixteenth century = 0.126 gm.
Sporta	In medieval Egypt = 500 ratl = 222.466 kg.
Stalln	Iron weight in Nassau (Germany). In Dillenburg = 160 pfund = 70.626 kg. In Siegen = 170 pfund = 79.5124 kg.
Stein, Stone	Market weight in Germany, Holland, Sweden, Poland; in different places of very different weight; e.g. in Amsterdam 3.952 kg; in Berlin the heavy stein = 10.285 kg, the light stein 5.142 kg; in Sweden = 13.556 kg; in Vienna = 11.202 kg. (See Niemann for further details.) In England the stone for fish or meat = 3.629 kg; the stone for glass = 2.268 kg; the stone for wool = 6.326 kg.
Surin, Condorin	Chinese = 10 li = $\frac{1}{10}$ tschen (tsien) = 1/100 liang (ling) = 1/1,600 catti = 1/160,000 pecul (pic) = 0.376 gm.
Tale, Tael, Tail, Tähl, Tao, Tole	Gold and silver weight in China, Japan, India; 1 tael = 10 mas (tsien) = 100 condorin (swin, fuen) = 1,000 cash (lei); 1 tael in China = 37.68 gm; Japan 37.6 gm; Java 68.75 gm; Siam 58.43 gm; Singapore 37.8 gm.

	In Sumatra 1 tal = $\frac{1}{8}$ bunkal = 1/100 catti = 2 pagodas = 16 mas = 64 copangs = 9.6 gm.
Tamino	See Madrid.
Tamuna	Islamic = 0.0147 gm.
Tan	See Pecul.
Tank	Indian, sixteenth century = 20.96 gm. See Bombay.
Tarry	Weight on Coromandel = 2 turkos = 12 seyras = 100 paloins = 6,000 pagodas = 3.336 kg.
Tasu, Tassug	Persian, 0.18 gm; since sixteenth century = 0.195 gm. Chinese gold weight = 10 jai = 100 miao = 1,000 mo = 10,000 tsiun = 1/10,000 li; 1 li = 0.05 gm.
Termino	Weight in Tunis for pearls and jewels; 1 termino = $\frac{1}{80}$ oncia = 0.4 gm.
Timbang	Weight in Batavia for grain and rice; 1 timbang = 2 amats = 7 kulack = 5 peculs = 10 sack = 295.360 kg.
Tipree, Tiprih	Weight for grain and rice in Bombay = $\frac{1}{2}$ seer = $\frac{1}{15}$ adowlie = 1/300 parah = 1/1,850 candy = 1/7,400 morah = 158.75 gm. In other places 1 tiprel = $\frac{1}{2}$ seer = $\frac{1}{48}$ adowlie = 1/128 parah = 1/1,024 candy.
Tod	Wool weight in England; 1 tod = $\frac{1}{4}$ wey = $\frac{1}{13}$ sack = 1/156 load = 2 stones = 4 cloves = 12.652 kg.
Tola	Gold and silver weight in East India: in Bengal and in Delhi 1 tola = 10.95 gm; in Bombay 11.56 gm; in Surat 14.4 gm; in Madras since 1811, 180 grains = 11.67 gm. In the sixteenth century 1 tola = 12 masa = 12.05 gm. According to Luther (1961) 1 tola = 11.664 gm = 12 masha = 96 rati; 1 rati = 8 chaval.
Tole, Tail	Gold weight in Manila = 270.17 gm; commercial weight = 297.17 gm.
Tomine	Gold and silver weight in Spain; 1 tomine = $\frac{1}{3}$ adarme = $\frac{1}{6}$ ochava = $\frac{1}{48}$ onça = 1/384 marc = 12 granos = 0.6 gm.
Ton, Tonne	Commercial weight of very different amounts at different places in Germany and other countries, also different for different merchandise. See Tun (Niemann p. 351).
Toque	Gold and silver weight in Pegu (Indochina), = 0.96 gm.
Trapasso, Trapeso	Southern Italy; on Malta 1 trapasso = $\frac{1}{32}$ oncia = 1/384 lira = 18 grains = 0.84 gm. In Naples and Sicily 1 trapasso = $\frac{1}{30}$ oncia = 1/360 libbra = 0.9 gm. See Palermo.

Tresseau	Old French pharmaceutical weight = 1 drachm or quentchen = 3.875 gm.
Troisquin, Troesken	Dutch = 2 deusquins = 0.2 gm.
Trouba-houache	Grain weight in Madagascar = 3.049 kg.
Troy weight, Troy mark	Old French market gold and silver weight. In England and Scotland 1 pound troy weight (also called troy mark) = 12 ounces = 240 pennyweight = 5,760 grains = 115,200 mites (see pound). The old French mark = $\frac{1}{2}$ troy pound = 8 ounces = 64 gros = 160 estelins = 192 deniers = 320 mailles = 640 felins = 4,608 grains (see marc) = 373.2 gm. In Holland 1 troy mark = 8 ounces = 160 engels = 640 vierlings = 1,280 troisquins = 2,560 deusquins = 5,210 ass.
Tschen, Tsien	Chinese commercial weight = $\frac{1}{10}$ leam (lyang) = 1/160 catti = 10 condorin (swin) = 100 li = 3.75 gm.
Tschopa, Tschupa	Rice weight in Sumatra = 456 gm.
Tsiun	Chinese gold weight = 10 sun = $\frac{1}{10}$ mo = 1/100 miao = 1/1,000 jai = 1/10,000 tchin = 1/100,000 fuen = 1/100,000,000 see; 100,000,000 tsiun = 1 li gold = 0.05 gm.
Tuko	See Tarry.
Tulam	India (Madras) = 144 paloins = 800 tolas = 9.331 kg (see Raju, 1962).
Tun, Tonne, Wey	In England the ton = 20 centner = 1,015.75 kg. Its weight is different in different places in England and also for different merchandise, e.g. the ton lead in London = 993.077 kg.
Uqiya	See Onça.
Usano, Ounce, Uncia, Unze	See Onça, Unze; = 2 loth. Islamic = $\frac{1}{12}$ ratl (see Hinz, pp. 39 ff.). Its weight differs in different places. In ancient Rome it was $\frac{1}{12}$ libra = 27.29 gm. In modern times in Cologne = 29.2 gm; in Amsterdam 30.9 gm; in Denmark 31.2 gm, etc. The pharmaceutical ounce in Amsterdam = 30.75 gm; in Venice 25.06 gm; in Vienna 35 gm; in Nuremberg 31.9 gm.
Vakea, Vaki	Abyssinia = 12 drachms = $\frac{1}{12}$ rottolo = 25.9 gm.
Vakega	See Mocha.
Vesno	Aleppo = $\frac{1}{7}$ cola = 5 rottoli = 3,000 drams = 9.5 kg.
Vierding	See Vienna; $\frac{1}{4}$ pound = 140 gm.
Viering	One quarter pound; called so in Nuremberg = 127.6 gm.

Vierting	Austrian name for $\frac{1}{4}$ pound or 8 loth = 140 gm.
Vis	See Madras; 1 vis = 5 seers = 120 tolas.
Vog	Russia and Norway; Norwegian vog = 36 pounds = 17.979 kg; Russian vog = 30 pounds = 14.980 kg.
Voll, Vall, Waal, Wall	Gold and silver weight in East India. In Bombay = $\frac{1}{40}$ tola = 1/960 seira = 0.291 gm. In Delhi 1 vall = $\frac{1}{32}$ tola = 0.36 gm.
Wag, Wage	Heavyweight for iron or lead. In France (vague) = 175 pounds = 85.665 kg. In Bruges = 79.980 kg; in Nassau (Germany) = 55.99 kg; in Saxony = 44 pounds in Sweden = 165 pounds = 69.85 kg.
Wageka	Gold and silver weight in Mocha = 10 caflas = 160 crats (karats) = 31.69 gm.
Wakea	See Vakea.
Weg, Wag	Danish = 3 bismar pounds = 36 Danish pounds = 17.975 kg.
Wey	English wool weight = $\frac{1}{24}$ last = $\frac{1}{2}$ sack = 6.5 tod = 13 stone = 26 cloves = 82.545 kg.
Wigge, Wigtje	Dutch = $\frac{1}{10}$ lood = 1/100 once = 1/1,000 pound = 1 gm.
Wiqr	See Harwar.
Xatague	East India (Coromandel) = $\frac{1}{18}$ seyra = 31.25 pagodas = 17.38 gm.
Yük	Islamic, a horse load = 162.144 kg. A silk yük in the sixteenth century = 61.5 kg.
Zentner	See Centner.
Zethim, Zethin, Satin	Since the thirteenth century in Hamburg and Lübeck = $\frac{1}{2}$ loth = $\frac{1}{8}$ vierting = $\frac{1}{16}$ halbmark = $\frac{1}{32}$ mark.
Zollpfund	See Berlin.
Zollzentner	See Berlin.
Zurlo	Syria (Aleppo) = 27.5 rottoli = 720 drams = 62.692 kg.

BIBLIOGRAPHICAL NOTES
AND BIBLIOGRAPHY

BIBLIOGRAPHICAL NOTES

In contrast to the extensive literature concerning metrology, the scientific publications about weighing instruments (weights and scales) are unfortunately scanty except for manuals of instruction on how to make balances.*

Lorenzo published a treatise in 1735 on the scales of ancient times, and this topic attracted various scholarly authors up to Kruhm's more literary booklet in 1934 on scales, Snyder's *Wagen und Waagen* (1957), and Häussermann's *Ewige Waage* (1962). Notable among these publications was Thomas Ibel's doctoral thesis on ancient and medieval scales, which was written with scholarly exactness and published in 1908. In 1908 also Ducros published a fully illustrated account of ancient Egyptian scales and the process of weighing in early Egypt (see our Fig. 4). In the same year a book on Egyptian weights and scales was published by Weigall, at that time an outstanding connoisseur in the field of Egyptian cultural history and archaeology. On this topic, however, the classic publications of Flinders Petrie remain the paramount work. His outstanding book on ancient weights and measures from excavations in Egypt appeared in 1926. A critical review by Glanville on weights and balances in early Egypt was published in 1935. A very valuable book, but out of print for decades, is Pernice's study of ancient Greek weights (1894).

The two books written by Decourdemanche (1909 and 1913) on ancient and modern weights and measures, especially those of the Near and Far East, contain nothing about the instruments used in weighing—that is, the weights and scales.

Only a few other examples of the literature on this topic can be mentioned. Ibel's doctoral thesis and the book of Sheppard and Musham have been quoted. Felibien's history of Paris contains an interesting report about the royal scales of this city (vol. 1, pp. 198–99). *A Short History of Weighing* by Sanders is instructive, as is, for a very limited field, Werner's booklet on scales and money in Merovingian times. Short, well-illustrated outlines of the history of scales and weights were published in 1924 by Dinse and in 1927 by Snyder. There is also the thorough volume by Alberti, *Mass und Gewicht* (1957), important in spite of its emphasis on metrology. A monograph, following an address, was published in 1958 by Ohnesorg. The history of weights in Holland is well covered by the book of Zevenboom and Wittop Koning and in papers

---

* See, for example, J. Leupold (1726) J. G. Leutmann (1729), and the *Istruzione* of Turin (1750). There are many others, including modern publications.

by Borssum-Buisman. The broad study by Testut (1946) deals mainly with the development of modern scales. It also contains much valuable historical information, as do Machabey's books (1949, 1962). The periodically published *News* of the Mettler Company in Switzerland and bulletins from Sartorius in Germany are very instructive about modern weighing machines.

Haeberlin's classic work on the Roman *as* (1910) cannot be regarded as a book on weights but, rather, on the development of Roman money and its origin from weight units. The same is true of most books on ancient coins, weights, and measures beginning with Budeus (1514) and continuing up to the most recent literature. An important exception is Professor Pink's publication (1938), which is a highly instructive scholarly work on Roman and Byzantine weights that have survived in Austria. Hinz (1955) is a valuable guide for old Islamic metrology, and there are many contributions in the new Indian journal *Metric Measures* concerning the metrology of ancient India.

The collection of medieval texts on weights critically compiled by Moody and Clagett (1952) is also of value.

Scholarly interest in scales and weights is also apparent in France. The important French publications, in addition to those by Machabey, include Charles Testut's book (1946), which contains a wealth of facts and a carefully compiled bibliography, and the books of Forien de Rochesnard and Lugan (1955), who have recently begun to publish a work on old French weights, more comprehensive than the previous worthy book by Gaillardie (1898) on the same topic. They have also published a book (Album No. 2) on African weights.

Collectors, scholars, and numismatists in the last century most frequently turned their attention to the field of money weights and gold scales. These once played an important role in national and international commerce and in daily life everywhere in Europe. A scholarly approach to this subject has been made by various authors; for example, Sheppard and Musham (1924) for England, Dieudonné (1925) for France, Mateu y Llopés for Spain (1934), and Colin Martin for Switzerland (1959). These instruments were used from antiquity to the mid-nineteenth century to check the weight of gold and silver coins. After the introduction of the metric-decimal system, the increased use of paper money, and the devaluation of gold and especially of silver, the need to weigh coins with special coin weights no longer existed.

Very little attention was formerly paid to the cultural history of weights and scales—their manufacture, the guilds of the weight makers and scale makers and their regulations, and the part that governmental decrees and ordinances have played in the history of these indispensable tools of commerce and science. Only recently has interest in this field claimed the attention of scholars. First among these is Charles Testut, whose work on the history of the French balance makers appeared in 1946. This was followed by books on the same subject by A. Machabey in 1949 and 1962. Martin Colin of Lausanne published in 1959 a very good history of the makers of gold scales in Geneva and Bern, and the present author's book on the scale and weight makers of Cologne and their guild was published in 1960(b). Cologne, like Antwerp, Amsterdam, Lyons, and Paris, has for centuries been a center of industry and commerce in Europe and consequently a center for scale and weight making.

In addition to these monographs on weighing, there is a great deal of scattered information on this topic in numismatic, archaeological, and antiquarian journals, in books on metrology, and so on. For many years I have collected such material, always aware that I could never hope to compile a complete bibliography, for that would be a full-time pursuit. However, all the known sources of valuable information about European scales and weights as tangible objects are presented in the bibliography that follows. The vast metrological literature is of course not included, and the weights and scales of the extra-European countries like China, India, South America, and even Russia could not be studied adequately. They are mentioned in the text only in passing.

BIBLIOGRAPHY

A. S-N. (‏א״ש-ן‎, Ephraim Stern). "Moznaim" (‏מאזנים‎, Scales), in *Enzyclopediah Mikraith*, *4*, Jerusalem, 1963, 540–43.

———— "Midoth umishkaloth" (‏מידות ומשקלות‎, Measures and Weights), ibid., pp. 846–78.

Adams, John Quincy. "Report upon Weights and Measures. Prepared in obedience to a Resolution of the Senate of the third March, 1817," Washington, 1821, 245 pp.

Adelung, Joh. Christoph. *Versuch eines vollständigen grammatisch-kritischen Wörterbuches der hochdeutschen Mundart*, *3*, Brünn, 1788.

Agricola, Georg. *Libri quinque de Mensuris et Ponderibus*, Basel, 1533.

———— *Berckwerk Buch* (German ed. by Philipp Bech), Frankfurt a/M, 1580.

Aiyar, B. S. D. "Reference, secondary and working standards," *Metric Measures*, *2* (6):15–24, 1958.

Alberti, Hans Joachin von. *Mass und Gewicht*, Berlin, 1957, 580 pp.

Alexander, J. H. *Universal Dictionary of Weights and Measures, Ancient and Modern; Reduced to the Standards of the United States of America*, 1st ed., Baltimore, 1850, 158 pp.; 2d ed., 1857.

Antonowitz, D. P. A., "Im Zeichen der Waage," *Rheinische Post*, 1957, No. 40.

Aubök, Jos. *Handlexicon über Munzen, Goldwerke . . .* , Vienna, 1893, 350 pp.

B.M.C. "Egyptian weights and balances," *Bull. of the Metropolitan Museum of Art*, pp. 85–90, 1917.

Bache, Alexander. "Report to the Treasury Department on the Progress of the Work of Constructing Standards of Weights and Measures and Balances in the Years 1846 and 1847," Exec. Doc. 73, 1–29, 30th Congress, 1st Sess., 1848.

Barbieux, Emil. "La Legislation française des poids et mesures," Doctoral thesis, Paris 1926, 201 pp.

Barnard, F. A. P. "The Metric System of Weights and Measures." An address delivered before the convocation of the University of the State of New York at Albany, New York, 1872; 194 pp.

Bernardus, Eduardus. *De Mensuris et Ponderibus Antiquis*, 2nd ed., Oxford, 1688, 347 pp.

Berriman, A. E. *Historical Metrology*, London and New York, 1953.

Beverinius, Bartholomaeus. *Syntagma de Ponderibus et Mensuris*, Lucca, 1711, 287 pp.

Beyerlinck, Laurentius. *Magnum Theatrum Vitae Humanae*, 8 vols., Leiden, 1665–66.

Bion, N. *Traité de la construction et des principaux usages des instruments mathematiques*, Paris, 1709.

Blaxland, S. G., and Bligh, E. W. *Sixty Centuries of Health and physick*, London [1931], xvi + 235 pp.

Block, Walter, *Messen und Wägen*, Leipzig, 1928.

Board of Trade. *Seventh Annual Report of the Warden of the Standards on the Proceedings and Business of the Standard Weights and Measures Department of the Board of Trade. For 1872–73*. Presented to both Houses of Parliament . . . XXII, London, 1873, 195 pp.

Borssum-Buisman, G. A. van. "Pijl of Sluitgewichten", *Jaarboek voor Munt-en Penningkunde*, *38*:94, 1951; and *39*:64–81, 1962.

───── "Over Munt Gewichten en Balansen," ibid., *40*:111–36, 1953.

Bosscha, J. "Relation des experiences qui ont servi à la construction de deux mètres étalons en Platin Iridié, comparés directement avec le mètre des Archives," *Ann. École Polytechnique Delft*, *2*:65–144, 1885.

Brandis, J. *Das Münz-Mass und Gewichtwesen in Vorderasien bis auf Alexander den Grossen*, Berlin, 1866.

Brauer, E. *Die Konstruktion der Waage*, Leipzig, 1906.

───── *The Construction of the Balance*, 3d rev. ed. by Fr. Lawaczeck. Trans. by H. Ch. Walters, published by the Incorporated Society of Inspectors of Weights and Measures, London, 1909, 314 pp.

Brenton, W. A. "Weighing Machines," in *Encylopedia Brittanica*, *23*: 479–86.

Brewer, Thomas. "On the antiquity of marking and stamping weights and measures," *J. Brit. Archeol. Ass.*, *8*:309–22, 1853.

Brisson, Mathurin-Jacques. *Instruction sur les mesures et poids anciens*, Paris, 1800, 129 pp.

Brøgger, A. W. "Ertog og Øre den gamb Norske vegt," *Videnskapsselskapets Skrifter II. Hist.-Filos. Klase*, 1921, No. 3 Kristiania 1921.

Budelius, R. *De Monetis et Re Numaria*, 2, Colonia Agrippinae, 1591.

Budeus, W. *De Asse et Partibus ejus*, Paris, 1514.

Burguburu, Paul. *Catalogue des poids anciens des villes de France*, Bordeaux, 1936.

Chambers, E. *Cyclopaedia or, an Universal Dictionary of Arts and Sciences*, London, 1728.

Chelius, J. K. *Zuverlassige Vergleichung sammtlicher Maase u. Gewichte der Haupstsadt, Frankfort a.M.*, Frankfort/Main, 1805.

Chevreul, "Examen critique de l'historie du mètre," *Comptes rendus Acad. Sci., Paris*, *69*:847–53, 1869.

Chisholm, H. W. *On the Science of Weighing and Measuring and Standards of Measures and Weight*, London, 1877, 192 pp.

Clark, Latimer. *A Dictionary of Metric and Other Useful Measures*, London 1891, 113 pp.

Clémenceau, E. *Le Service des poids et mesures en France à travers les siècles*, Saint-Marcellin-Isère, 1909.

Colin, Martin. Les Boîtes de changeurs à Genève et à Berne (XVIIe–XVIIIes.), *Rev. suisse Numismat.*, *39*:59–109, 1959.

Commissionsbericht. See Oesterreich.

Conservatoire National des Art et Métiers. *Catalogue du Musée*, Section K, "Poids et mesures métrologie," Paris, 1941.

"Convention du mètre signée le 20 Mai 1875," Paris, Imprimeur-Libraire du Bureau des Longitudes, de l'École Polytechnique, 1875, 17 pp.

Cumberland, Rich. *An Essay towards the Recovery of the Jewish Measures & Weights*, London, 1686, 140 pp.

Dean. See Epiphanias.

Decourdemanche, J. A. *Traité pratique des poids et mesures des peuples anciens et des Arabes*, Paris, 1909, 144 pp.

——— *Traité des monnaies, mesures et poids anciens et modernes de l'Inde et de la Chine*, Publication de l'Institute ethnographique International de Paris, Paris, 1913, 172 pp.

Delamorinière et Séguir. "Projet d'une nouvelle form de poids, depuis celui de cinquante kilogrammes jusqu'a celui d'un gramme," *Comptes rendus Acad. Sci.*, Paris, *44*:531–33, 1857.

Depping, G. B. *Les Reglemens sur les arts et metiers de Paris*, Paris, 1837, 474 pp.

Desaguliers, J. T. "An experiment to compare the Paris weights as they are now used at Paris with the English weights," *Phil. Trans. Roy. Soc. London*, *31*:112, 1720.

Desai, H. N. "Enforcement of metric system in Gujerat," *Metric Measures*, 6(6):18–20, 1963.

Deschamps de Pas, L. "Note sur quelques poids monetaires," *Revue Numismat.* (n.s.) *8*:270–78, 1863.

Dhurandhar, B. R. "Working standard balances; Part 3: Beam design," *Metric Measures*, 6(4):10–14, 1963.

——— "Working standard balances; Part 4: Knife-edges," *Metric Measures*, 7(2):13–15, 1964.

——— "Ten years of weights and measures division in National Physical Laboratory of India," ibid., pp. 16–20.

Diderot, Denis. *Encyclopédie*, 3d ed., *4*, Geneva, 1778.

Dietrich, Albert, *Zum Drogenhandel im islamischen Aegypten*, Heidelberg, 1954.

Dieudonné, A. *Manuel des poids monétaires*, Paris, 1925, 184 pp.

Dinse, E. *Fortschritte im Waagenbau*, Berlin, 1924.

Diringer, David. "Le Iscrizioni antico-ebraiche Palestinesi," *Publ. R' Univers. Firenze*, 3 Ser., *2*, Florence, 1934.

Doursther, Horace, *Dictionnaire universel des poids et mesures anciens et modernes*, Brussels, 1840.

Ducros, Hippolyte. "Étude sur les balances Égyptiennes," *Ann. du Service des Antiquités de l'Égypt*, *9*:32–53, 1908.

———— "Deuxième étude sur les balances Égyptiennes," ibid., *10*:240–53,1910.

Dumas, Alexandre. "Système métrique," *Comptes rendus Acad. Sci.*, Paris, *69*:514–18, 1869.

Dwivedi, R. N. "Enforcement of metric system among Adibasis," *Metric Measures*, *6*(6):12–14, 1963.

Ebers, Georg. "Papyrus Ebers I. Die Gewichte und Hohlmaasse des Papyrus Ebers," *Abt. Kgl. Sächs. Ges. Wiss. Philolog. Hist. Kl. II.* Leipzig, 1889.

Eisenschmid, Jo. Casp. *De Ponderibus et Mensuris veterum Romanorum, Graecorum, Hebraeorum, Argentorati*, Strassbourg, 1737.

Epiphanias (Dean, James Elmer). *Epiphanias' Treatise on Weights and Measures. The Syrian Version*, Chicago, 1935.

Ercker, Lazarus. *Beschreibung aller fürnemsten Mineralischen Ertzt und Berckwerksarten*, Prague, 1574. English trans. by Anneliese Grünhaldt Sisco and Cyril St. Smith, Chicago, 1951.

Evelyn, Sir George Shuckburgh. "An account of some Endeavours to ascertain a Standard of Weights and Measure," *Phil. Trans. Roy. Soc.*, London, *88*:133–82, 1798.

Fahmy, Abdel Rahman. *Early Islamic Coinweights*, Cairo, 1957.

Falck-Muns, Rolf. "En middelalders vektlodd fra Nordtrondelag," *Nordtrondelags historidags arbok*, pp. 89–92, 1939.

Felibien, Michel. *Histoire de la ville de Paris*, 5 vols., Paris, 1725.

Ferrand, Gabriel. "Les poids, mesures et monnaies des mers de sud aux XVIᵉ et XVIIᵉ siècles," extrait du *Journal Asiatique*, 1920; Paris, 1921, 269 pp.

Forien de Rochesnard, Jean, and Jacques Lugan. *Catalogue général des poids*, Anvers, 1955.

———— *Album No. 1: Poids Français*, Anvers, 1957, 225 pp.

———— *Album No. 2: Poids d'Afrique*, Anvers, n.d. [1959], 296 pp.

Fossati, Spiritu. *De Ratione nummorum, Ponderum et mensurarum in Gallis*, Turin, 1842, 145 pp.

France. Bureau de Verification des poids et Mesures, *Rapport sur la révision des étalons des bureaux . . . de l'Empire francais en 1867 et 1868*, Paris, 1871, 63 pp.

Fresenius, Phil. *Das Grammengewicht und seine Anwendung in der ärztlichen Praxis*, Frankfurt a/M., 1868; 2d ed., 1889.

Fryer, John. *A New Account of East India and Persia*, London, 1698.

G.V. *Tavole di confronto delle misure piacentine colle misure del nuovo sistema metrico*, Piacenza, 1840, 404 pp.

Gaillardie, Louis. *Poids anciens des villes de France*, Paris, 1898, 57 pp.

Galen, Cl. *De Compositione Medicamentorum, Eiusdem de Ponderibus et Mensuris*. Trans. by Joannes Guinterus Andernacus, Basel, 1530.

——— *Les troys premiers Livres de Claude Galien de la Composition des Medicamens en General*, Tours, 1545.

Garcia, Jos. *Breve Contego, y Valance de las pesas y Medidas*, Madrid, 1731.

Gardiner, Sir Alan. *Egypt of the Pharoahs*, Oxford, 1961, 461 pp.

Garrault, François. "Reduction et avalvation des mesures et poids anciens du duché de Rethelois à mesures et poids royaux. Mise & rédigée par escrit en presence des deputez du dit Duché." Paris, 1585.

Georges, K. E. *Ausführliches lateinisch–deutsches Handwörterbuch*, 2 vols., Leipzig, 1880.

Glanville, S. R. K. "Weights and balances in ancient Egypt," *Proc. Roy. Inst. of Great Britain*, *29*:10–40, 1937.

——— "Weights and balances in Ancient Egypt," *Nature*, *137*: 890–92, 1936.

Gotz, Ernst. *Waagen und Wiegeeinrichtungen*, Leipzig, 1931.

Gregory, G. *A New and Complete Dictionary of Arts and Sciences*, 3 vols., New York, 1822.

Griffenhagen, George, "Pharmacy Museums," Madison, Wis., 1956.

——— "Tools of the Apothecary," Washington, 1957.

Griffith, F. L., "Notes on Egyptian weights and measures," *Proc. Soc. Biblical Archaeol.*, *14*:403–40, 1892; *15*:301–15, 1893.

Guerra, Francisco. "Weights and measures in pre-Columbian America," *J. Hist. Medicine*, *15*:342–44, 1960.

Guichard, M. *De la Sensation à la méthode de mesure*, Paris, 1937.

Guilhiermoz, P. *Remarques diverses sur les poids et mesures du moyen age*, Bibliotheque de l'Ecole des Chartes, Paris, *80*:1–100, 1919.

Guillemot, A., Papavoine, and Moraut. *Renseignments sur la service, la verification et la fabrication des poids et mesures*, Chalons-sur-Marne, 1902, 647 pp.

Gurley's Handbook of Weights and Measures for the Use of Scalers, 4th ed., New York, 1912, 202 pp.

Haeberlin, E. J. *Aes Grave. Das Schwergeld Roms und Mittelitaliens*, Frankfurt a/M., 1910.

Häussermann, Ulrich. *Ewige Waage*, Cologne, 1962, 90 pp.

Hallock, W., and H. T. Wade. *Outlines of the Evolution of Weights and Measures and the Metric System*, New York, 1906.

Halsey, F. A. *The Metric Fallacy*, 2d ed., New York, 1920, 229 pp.

Harkness, William. "The progress of science as exemplified in the art of weighing and measuring," Presidential address, *Bull. Phil. Soc. Washington, 10*: 39–86, 1888.

Harris, John. *Lexicon Technicum or an Universal English Dictionary of Arts and Sciences*, London, 1723.

Hartlaub, G. F. "Die Nürnberger Waage," *Die BASF, 4*:157–58, 1954.

Hassler, F. R. "Instructions Relating to the Use of Standard Weights," House Doc. 454, 25th Congress, 2d Sess., 1839.

Hillinger, Bruno. "Studien zu mittelalterlichen Massen und Gewichten," in *Seeligers Histor. Vierteljahrsschrift, 3*: 161–215, 1900.

Hinz, Walther. *Islamische Masse und Gewichte. Handbuch der Orientalistik Erg, 1*, Part 1, Leiden, 1955.

Holzmair, Ed. "Die Gewichtssammlung des weiner Münzkabinettes," *Numismat. Zsch. Wien, 65*:99–106, 1932.

Hoops, J. *Reallexicon der germanischen Altertumskunde*, Strassburg, 1918/19.

Hornbostel, E. von. "Die Herkunft der altperuanischen Gewichtsnorm," *Anthropos, 26*:255–58, 1931.

Hostus Matthaeus. *Historia antiquitatis Rei nummariae Mensurarum, Ponderum etc.*, Rechenberg ed., 3 vols., Amsterdam, 1692 and 1698.

Hultsch, Friedr. *Griechische und römische Metrologie*, Berlin, 1862.

——— *Metrologicorum Scriptorum Reliquiae*, 2 vols., Leipzig, 1864, 1866.

——— "Die Gewichte des Altertums nach ihrem Zusammenhange dargestellt," *Abt. Kgl. Sächs. Ges. Wiss. Philolog. Hist. Kl. XVIII*, Part 2, 1898, 205 pp.

Ibel, Thomas. *Die Wage bei den Alten*, Program des K. Luitpoldprogymnasiums, Forchheim, 1906.

——— "Die Wage im Altertum und Mittelalter," Doctoral thesis, Erlangen, 1908.

Ingalls, W. R. *Modern Weights and Measures*, American Institute of Weights and Measures, New York, 1937.

Instruction sur les mesures déductes de la grandeur de la terre ... , 2d ed., Angers, An II (1794).

Irwin, Keith Gordon. *The Romance of Weights and Measures*, New York, 1960, 144 pp.

Istruzione per li fabbricatori, et agquistatori delle bilance, stadere e misure, estesa d'ordine dell excellentissima camera a tutte provincie del Piemonte, Turin, 1750.

Jansson, S. O. *Måttordbok*, Stockholm, 1950, pp. 13–14.

Jaubert. *Dictionnaire raisonné universal des arts et metiers*, Paris, 1773.

Jüthner, Julius. "Examen," *Jahreshefte des österr. archeol. Inst. Wien*, 16: 1913; Beiblatt, Spalte, 197–206.

Juncker, Johann. *Conspectus Chemiae Theoretico-Practicae. Vollständige Abhandlung der Chemie*, Halle, 1749.

Kanwarjit Singh. "Records and returns for weights and measures organisation," *Metric Measures*, *6*(6):15–17, 1963.

Karpin, E. B. *Wägemaschinen*, Leipzig, 1960.

Kelemen, Pál. *Mediaeval American Art*, New York, 1943.

Kisch, Bruno. "Two remarkable Roman stone weights in the Edward C. Streeter Collection at the Yale Medical Library," *J. Hist. Medicine*, *9*:97–100, 1956.

——— "Weights and scales in mediaeval Scandinavia," *J. Hist. Medicine*, *14*:160–68, 1959.

——— "The medical use of scales," *Amer. J. Cardiol. 5*:202, 1960a.

——— *Gewichte und Waagemacher im alten Köln* (*16–19 Jahrhundert*), Köln, 1960b [1962], 180 pp.

Koehnle, Joh. *Handbuch für den praktischen Waagenbau*, Bielefeld, 1910.

Kohlbusch, Herman. *Illustrated Price List*, New York, 1902, 38 pp.

Kretzenbacher, Leopold. *Die Seelenwaage*, Klagenfurt, 1958, 243 pp.

Kruhm, August. *Die Waage im Wandel der Zeiten*, Frankfurt a/M, 1934.

Kubitschek, J. W. "Gewichtstücke aus Dalmatien," *Archeol.-Epigraph. Mitteilungen aus Oesterreich-Ungarn, 15*: 85–90, 1892.

Küntzel, Georg. *Über Verwaltung des Mass-und Gewichtswesens in Deutschland während des Mittelalters*, Leipzig, 1894.

Kunz, G. F. "The new international metric carat of 200 milligrams," *Trans. Amer. Inst. Mining Engineers* (New York Meeting, Feb. 1913), pp. 1225–45.

——— "The international language of weights and measures," *Scientific Monthly, 4*:215–19, 1917.

La Condamine, de. "Nouveau projet d'une mesure invariable propre à servire de mesure commune à toutes les Nations," *Mém. Acad. Roy. Sci.* pp. 489–514, 1747.

Labielle. *Cours des Poids et Mesures*, Paris, 1932.

Laignel-Lavastine. *Histoire Générale de la Medicine*, vol. 1, Paris, 1936.

La-Ramée Pertinchampt, Bidone, Vassati-Eaudi, Merlini, M. X. Provana, *Tables de comparison entre les poids et mesures du nouveau système et les poids et mesures ci-devant en usage à Turin*, Turin, 1809, 415 pp.

Layard, Austen H. *Discoveries in the Ruins of Nineveh and Babylon*, London, 1853.

——— *Nineveh and Babylon*, London, 1874.

Lazzarini, M. "Le Bilance romane del museo nazionale e dell antiquarium comunale di Roma," *Atti della Accademia Naz. dei Lincei*, 8 Ser., *3*:221–54, 1948.

Leake, Chauncey, D. *The Old Egyptian Medical Papyri*, Kansas, 1952.

Le Blanc. *Traité historique de monnoyes de France*, Paris, 1703.

Lemale, A. G. *Monnaies, poids, mesures et usages commerciant des tous les etats du Monde*, 2d ed. Paris, 1875, 108 pp.

Leupold, Jacob. *Theatrum Staticum Universale I. Das ist Schauplatz der Gewicht-Kunst und Waagen*, Leipzig, 1726.

Leutmann, Joh. Georg. "De Bilancibus et Novis Inventis Staticis," *Comment. Acad. Scientiar. Imper. Petropolitanae*, *2*:35–81 (1927), 1729.

Levy, Reuben. *The Ma'alim al Ourba Flakham al Hisba*, London, 1938.

Lindquist, Sune. *Fran Upplands Forntid*, Uppsala, 1956.

Lorenzo, Conte Luigi. Dissertatione IX, "Sopra de Bilancie degli Antichi," *Saggi di Dissert. Accad. Publicamente lette nella Nobile Academia Etrusca*, *3*:93–102, 1735.

Lunier, M. *Dictionaire des sciences et des arts*, Paris, 1806.

Luther, M. R. "Conversion tables for tola and its fractions," *Metric Measures*, *4*(6):18–23, 1961.

Macfarlane, John J. *Conversion-Tables*, 10th ed., Philadelphia, 1929, 109 pp.

Machabey, Armand J. *Mémoire sur l'historie de la balance et de la balancerie*, Paris, 1949.

——— *Poids et mesures du Languedoc et des provinces voisines*, Toulouse, 1953, 143 pp.

——— "La Metrologie dans les musées de Provence," Sorbonne, 1959.

——— "Les progrès de la précision des mesures," *J. Hist. Medicine*, *15*: 372–83, 1960.

——— "Historie des poids et mesures depuis le 13 siècle. La metrologie dans les musées de Provence." Thèse pour le docteur et de l'Université soutenue en Sorbonne le 19 Juin 1959, Paris, 1962, 512 pp.

Mackay, Ernest, J. H. *Chanhu-Daro Excavations 1935–36*, New Haven, Conn., 1943.

Mariana, Joh. *De Ponderibus et Mensuris*, Mainz, 1605, 160 pp.

Marshall, Sir John. *Taxilla*, Cambridge, 1957.

Martin, Saint-Léon, E. *Historie des corporations de metiers*, 3d ed., Paris, 1922, 876 pp.

Martin, William. *An Attempt to Establish Throughout his Majesty's Dominions an Universal Weight and Measure Dependant on Each Other*, London, 1794, 39 pp.

Massarius, D. *De Ponderibus et Mensuris Medicinalibus Libri Tres*, Zurich, 1584, 107 pp.

Mateu y Llopes, F. *Catálogo de los ponderales monetarios del museo arqueologico nacional*, Madrid, 1934, 290 pp.

Mettler, *News*, Nos. 1–29, no year, no city [U.S.A.].

Monumenta Germaniae: Historica: Leges: Constitutiones, vol. 1, 1893.

Moody, E. A. and Marshall Claggett. *The Medieval Science of Weights*, Madison, 1952.

[Morin, A.] "Notice historique sur le système métrique sur ses dévcloppmenets et sur sa propagation," *Ann. Conserv. Arts. et Metiers*, Paris, 1873.

Muller, John. *Indian Tables for the Conversion of Indian Mun*, Calcutta, 1836, 294 pp.

National Industrial Conference Board. Research Report No. 42, October 1921, *The Metric versus the English System of Weights and Measures*, New York, 1921, 261 pp.

Nelkenbrecher, J. C. *Allgemeines Taschenbuch der Munz-, Maass-, und Gewichtskunde*, 17th ed., Berlin, 1848, 206 pp.

Nicholson, Edw. *The Story of Our Weights and Measures*, Liverpool, 1901.

—— *Men and Measures. A History of Weights and Measures Ancient and Modern*, London, 1912, 313 pp.

Niemann, Fr. Alb. *Vollständiges Handbuch der Münzen, Masse und Gewichte aller Länder der Erde*, Quedlinburg and Leipzig, 1830.

Nissen, Heinrich. "Griechische und römische Metrologie," in *Handbuch der classischen Altertumswissenschaft*, ed. Ivan Müller, 2d ed., Nordlingen, 1886, pp. 835–90.

Noback, Friedrich. Münz- Maass und Gewichtsbuch, 2d ed., Leipzig, 1877, 1166 pp.

Norris, E. "On the Assyrian and Babylonian weights," *Roy. Assyriol. Soc.*, *16*:1856.

Nowotny, Eduard. "Zur Mechanik der antiken Wage," *Jahreshefte des österr. archaeol. Inst. Wien, 16*:1913; Beiblatt, Spalte 5–36.

Oehlschläger, J. C. "Handbuch für Kaufleute," in Ch. F. Grieb, *Englisch-Deutsches und Deutsch-Englisches Worterbuch, 2*, Philadelphia, 1857.

Oesterreich. K. K. Handels-Ministerium, *Commissionsbericht über das Verhältnis des Bergkrystall-Kilogramms welches bei Einführung des metrischen Maasses und Gewichtes das Urgewicht in Österreich bilden soll zum Kilogramm der Kaiserlichen Archive in Paris*, Wien 1870, 136 pp.

Ohnesorg, Wolfgang von. *Von Waagen und Waagmachern*, Frankfurt a/M, 1958.

Olaus, Magnus. *Historia de Gentibus Septentrionalibus*, Rome, 1555.

Ouin-Lacroix, Ch. *Histoire des anciennes corporations et métiers et des confrèriers religieuses de la capitale de la Normandie*, Rouen, 1850, 761 pp.

Owen, George, A. *A Treatise on Weighing Machines*, London, 1922, 202 pp.

Paetus, Lucas. *De Mensuris et Ponderibus Romanis et Graecis*, Venice, 1573.

[Paucton.] *Métrologie, ou traité des mesures, poids et monnoies des ancients peuples & des moderns*, Paris, 1780, 955 pp.

Paulitschke. *Beiträge zur Ethnographie und Anthropologie der Somal, Galla und Harari*, Leipzig, 1886.

Pauly. *Realenzyclopedie der classischen Altertumswissenschaft*, Neue Bearbeitung, vol. 22, Stuttgart, 1952–61.

Pednekar, V. M. "How weights and measures are enforced in Bombay State," *Metric Measures, 1*(2):27–33, 1958.

Pennsylvania Senate. "Report relative to weights and measures," Harrisburg, 1838, 6 pp.

Pernice, Erich. *Griechische Gewichte*, Berlin, 1894.

Petrie, Flinders, "Glass weights," *Numismat. Chron.* 4 Ser., *18*:115, 1918.

—— *Ancient Weights and Measures*, London, 1926.

—— *Measures and Weights*, London, 1934.

—— *Ancient Gaza IV. Tel El Ajjid*, London, 1934.

Pharmacopoeia Matritensis, 2d ed., Madrid, 1762.

Pink, Karl. "Römische und byzantinische Gewichte in österreichischen Sammlungen," *Sonderschr. des österr. atchaeol. Inst. Wien, 12*: 1938.

Pinkerton, John. *An Essay on Medals*, London, 1789.

Putnam, J. Pickering. *The Metric System of Weights and Measures*, Boston, 1877, 83 pp.

277

Queipo, V. Vazquez. *Essai sur les systèmes metriques et monetaires des anciens peuples*, 4 vols., Paris, 1958.

Raju, L. "Development of weights in Madras area: Part 1," *Metric Measures*, 5(1):11–16, 1962.
——— "Development of weights in Madras area: Part 2," ibid. (2), pp. 3–9.
Rapport sur la révision des étalons, Paris, 1871.
Rây, P., ed. *History of Chemistry in Ancient and Medieval India*, Calcutta, 1946.
Rayleigh, Lord, and William Ramsay. *Argon, A New Constituent of the Atmosphere*, Washington, 1896, 43 pp.
Reifenberg, A. "Ein neues hebraisches Gewicht," *J. Palestine Oriental Soc.*, 16:39–43, 1936.
Research Report. See National Industrial Conference B.
Rhodius, Jo. *De Ponteribus et Mensuris Veterum Medicorum*, Copenhagen, 1672.
Ridgeway, William. *The Origin of Mettalic Currency and Weight Standards*, Cambridge, 1892.
Roberval, M. de. "Nouvelle manière de balance," *Mém. Acad. Roy. Sci.*, 10:494–96, 1730.

Sabatier, J. "Lettre aux directeurs de la Revue Numismatique. Poids byzantins de cuivre," *Revue Numismat.* (n.s.) 8 : 6–18, 1863.
Sachtleben, Rud. "Von alten Waagen," *Die BASF*, 4:159–68, 1954.
Sanders, L. *A Short History of Weighing*, Birmingham, 1947, 59 pp.
Santorio, Santorio. *De Statica Medicina*, Venice, 1614.
Sartorius. "Electronische Microwaage 4101," [Göttingen, 1961].
Schätze aus Peru. Catalogue of the 1959 exhibit in the Rautenstrauch-Joest Museum in Cologne; Recklinghausen, 1959.
Schmelzer, K. "Zur Entwicklung der Brückenwaage," *Glasers Ann.* 64:27–32, 1940.
Scott, R. B. Y. "Weights and measures of the Bible," *Biblical Archaeol.*, 22:22–40, 1959.
Sedláček, August. "Paměti a doklady o staroćeských mirách a váhách" (Reminiscences and documents of old Bohemian measures and weights), *Rozpravy ćeské Avad. véd aumněn*, Class 1, No. 66, Prague, 1923.
Segré, Angelo. *Metrologia e circolazione monetaria degli antichi*, Bologne, 1928.
Sharp, Ogen, W. "The Roman mint and early Britain," *Brit. Numismat. J.*, 1 Ser., 5:1–50, 1909.
Sheppard, T., and J. F. Musham. *Money Scales and Weights*, London, 1924.

Sigaud de la Fond. *Description et usage d'un cabinet de physique experimentale*, vols. 1 and 2, Paris, 1775.

Simienowicz, Casimir. *Vollkommene Geschütz-Feuerwerck- und Buchsen-meisterey-Kunst*, Frankfurt a/M, 1676.

Smith, Ralph W. "The federal basis for weights and measures," Nat. Bureau of Standards, Circular 593, Washington, 1958.

Snyder, Geerto. *Wägen und Waagen*, Ingelheim, 1957, 62 pp.

Sökeland, Herrman. "On ancient desemers or steelyards," *Smithsonian Reports*, p. 551–64, 1900.

Soloweitschik, Max. *Die Welt der Bibel*, Berlin, 1926, 240 pp.

Steinheil, C. A. von. "Über das Bergkrystall-Kilogramm auf welchem die Feststellung des bayrischen Pfundes nach der Allerhöchsten Verordnung vom 28. Februar 1809 beruht," *Kgl. Bayer. Akad. Wiss.*, 4: I Abt., 1837.

———— Über genaue und invariable Copien des Kilogrammes und des Mètre, prototype der Archive zu Paris welche in Österreich bei Einführung des metrischen Maass- und Gewichts-systems als Normaleinheiten dienen sollen und über die Mittel zu ihrer Verfielfältigung," *Denkschr. Akad. Wiss. Wien*, 27:1867.

Stengel, Walter. "Die Merkzeichen der Nürnberger Goldschmiede," *Mitteilungen aus dem germanischen Nationalmuseum Nürnberg*, 1915/1918, pp. 107–55.

Stern, Ephraim. See A.S-N.

Stillhard, M. *Units—Masseinheiten—Unité de mesure*, Basel–New York, 1963.

Tacchi, G. B. L. *Manuale di Metrologia*, Rovereto, 1876, 395 pp.

Tarbé, S. A. *Manuel pratique et élémentaire des poids et mesures et du calcul decimal*, 5th ed., Paris, An XII (1803).

Taylor, Alfred B. *Octonary Numeration and its Application to a System of Weights and Measures*, Philadelphia, 1887, 73 pp.

Testut, Charles. *Memento du pesage; Des instruments de pesage: Leur histoire à travers les ages*, Paris, 1946.

Thorwald, Jürgen. *Macht und Geheimnis der frühen Aerzte.* Munich-Zurich, 1962, 331 pp.

Thureau-Daugin, F. *Textes mathématiques Babyloniens*, Leiden, 1938, xl +243 pp.

Thurston, Robert H. *Conversion Table of Metric and British or United States Weights and Measures*, New York, 1883, 83 pp.

Tufnell, Olga. *Lachisch III. The Iron Age*, London, New York, Toronto, 1953.

Ulrich-Bansa, Oscar. *Moneta Mediolanensis*, Venice, 1949.

279

United States Department of State. "Report upon weights and measures"; see Adams, J. Q.

Vazquez, Queipo, V. See under Q.

Viedebantt, Oscar. *Zur Metrologie des Altertums*, Leipzig, 1917.

—— *Antike Gewichtsnormen und Münzfüsse*, Berlin, 1923.

Vieweg, Richard. "Aus der Kulturgeschichte des Messens," *Stahl und Eisen*, *80*:265–72, 1960.

—— *Mass und Messen in kulturgeschichtlicher Sicht*, Wiesbaden, 1962, 28 pp.

Walker, J. "Some recent oriental coin acquisitions of the British Museum," *Numismat. Chron.*, Part 4, Ser. 5, No. 60, pp. 241–53, 1935.

Wallis, Joh. *Mechanica, Sive de Motu, Tractatus Geometricus*, London, 1670.

Weigall, A.[rthur] E. [P.] "Some Egyptian weights in Prof. Petrie's collection," *Proc. Soc. Biblical Archaeol. 33*:378–95, 1901.

—— *Catalogue général des antiquités Egyptiennes du Musée du Caire. No. 31271–31670. Weights and Balances*, Cairo, 1908.

Weigel, Christoff. *Abbildung der gemeinnützlichen haupt Stände*, Regensburg, 1698.

Werner, Joachim. "Waage und Geld in der Merowingerzeit," *Sitz-Ber. Bayer. Akad. Wiss. Philos. Hist. Kl.* Part 1, Munich, 1954.

Wideen, Harold, "En bronshäst från medeltiden," *Goteborg's Historiska Museum årstryck*, pp. 18–29, 1953.

—— "Bronshästen från Vestenhaga," *Hilten-Cavalliusföreningens årsbok*, pp. 29–36, 1954.

Wild, Michael Fried. *Uber allgemeines Maas und Gewicht*, 2 vols., Freyburg, 1809.

Witte, Alphonse de. "Quelques ajusteurs jurés des ponds et balances en functions aux Pays-Bas autrichiens," *Rev. Belge Numismat.*, pp. 49–94, 1895.

—— *Poids de marchandises des anciennes provinces belgiques*, Brussels, 1890.

Wolf, M. C. "Recherches historiques sur les étalons de poids et mesures de l'observatoire et les appareils qui ont servi à les construire," *Ann. Observatoire de Paris, Memoires, 27*: 1882.

Wolff, Philippe. *Note sur les poids et mesures employés à Toulouse aux XIVe et XVe siècles dans commerces et marchands de Toulouse vers 1350–1450*, Paris, 1953.

Wrede, Fab. "Note sur le mètre et le kilogramme," *Bihangtill K. Svenska vet. Akad. Handlingar, 1*(3), 1–40 (of the reprint), 1872.

Wurtele, Arthur. *Standard Measures of the United States, Great Britain and France*, New York and London, 1882.

Yadin, Yigael. "Ancient Judaean weights and the date of the Samarian ostraca. Studies in the Bible," *Scripta Hierosolymitana*, 8:1–17, 1960.

Zedler. *Grosses vollständiges Universallexikon aller Wissenschaften und Künste*, Halle and Leipzig, 1731–54.

Zevenboom, K. M. C., and D. A. Wittop Koning. *Nederlandse Gewichten*, Leiden, 1953.

INDEX OF NAMES AND PLACES

Appendices 2 and 3, which are alphabetical, are not indexed.

289

SUBJECT INDEX

Appendices 2 and 3, which are alphabetical, are not indexed.

International Conference on Weights and Scales, 91
Iron scale, 12, 47
Ivory, 33, 47, 61, 65

Jewels, Jewelers, 33, 52, 60, 65, 79, 137, 145 ff., 163
Jugum, 32
Juno Moneta, 48
Justice, 78

Karate, Kirat, 145
Kedet, Kite, Kat, 8, 147
Keration, 145
Kikar, 11
Kilogramme: des Archives, 16 ff., 84, 90; définitive, 17; provisoire, 17, 18, 83
Kirschenwaage, 12
Knife-edge, 36, 42, 48
Knob, on weights, 102
Knopf, 102
Knucklebone. *See* Astragalus
Koran, 8
Kornwaage, 54, 55
Krämerwaage, 55
Krautwaage, 13

Laden, 137, 172 ff.
Lanx, 27, 211
Laufgewicht, 56
Lapis lazuli, 1
Laws, re weights and scales, 4–6, 9, 12–14, 18 ff., 28, 33, 48, 58, 69, 76, 83, 84, 137, 146, 164, 211
Libra, 6, 11, 17, 27, 152
Lingula, 36
Lion weights, 116, 118
Lira, Litra. *See* Libra

Machina anthropometrica, 76
Magna Charta, 14
Manipulus, 1
Manufacture, scales and weights, 165, 211
Mark, *Marc*, 5, 6, 9, 16, 109, 126, 127
Market: scales, 33; weights, 6, 83, 84, 103, 148, 154, 163
Marmorarii, 167
Masse de compensation, 56
Master cup, nested weights, 126 ff.
Mastersigns: 5, 41, 50, 63, 70, 98, 107, 127, 134, 137–39, 163, 167 ff., 213; Antwerp, 190 ff.; boxmakers, 164; Cologne, 170 ff.; coppersmiths, 174; Nuremberg masters, 184 ff.; nested weights, 175 ff.

Mechanics of weighing, 34, 58 ff., 60
Medals, commemorative, 18, 21–23, 77
Medicine, 5, 7, 65, 66
Medieval era: Eastern weights, 97, 98, 106, 108, 131, 152; French weights, 103, 155 ff.; guilds, 167; literature re weighing, 265, 266; money scales, 69; Peruvian art, scales in, 34; pharmaceutical weights, 97, 98; regulations re weighing, 4, 5, 9, 12, 13
Mercury, 85, 89
Mermaids, 129
Messing, 90
Mètre des Archives, 16 ff.
Mètre provisoire, 18
Metric-decimal system, 11, 15 ff., 76, 80, 90, 95, 141, 266; History of, 18 ff.; nomenclature, 25
Metrologic dictionary, 58
Metrology, 3, 14, 15 ff., 140 ff., 150, 265, 266
Microscales, 26, 56
Mintmaster, 9
Mints, 6, 7, 52, 83, 126, 130, 211; symbols of French, 162
Mna, Mina, 97, 146
Moneta usualis, 154
Monetiform weights, 4, 103, 145, 195
Money: changers, 9, 70, 83, 130, 134, 137, 163; scales, 65, 69 ff., 137; weights, 70, 129, 154, 266
Monograms on weights, 154
Monopoly: of coinage, 4; weights, 126, 213
Mother-of-pearl, 47
Münzgewichte, 129 ff.
Münzordnung, 9
Muhtasib, 11, 163
Muttergewichte, 6, 88

Nesef, 93
Nested weights, 83, 122 ff., 126 ff., 129, 163, 168, 175 ff., 212
Nickel–chromium, 89
Nomisma, 70, 153 ff., 222
Normal-Aichungscommission, 90
Notiometra, 55
Nub, 147

"Opium" scales, 33, 61, 65
Ornamentation, weights, 113, 146 ff., 154

Pans, of scales, 27, 36, 41, 42, 47 ff., 55, 70 ff., 74
Papyrus, 1, 26
Passiergewicht, 130